Biosemiotics

Volume 18

Aims and Scope of the Series

Combining research approaches from biology, semiotics, philosophy and linguistics, the field of biosemiotics studies semiotic processes as they occur in and among living systems. This has important implications and applications for issues ranging from natural selection to animal behaviour and human psychology, leaving biosemiotics at the cutting edge of the research on the fundamentals of life.

The Springer book series *Biosemiotics* draws together contributions from leading scholars in international biosemiotics, producing an unparalleled series that will appeal to all those interested in the origins and evolution of life, including molecular and evolutionary biologists, ecologists, anthropologists, psychologists, philosophers and historians of science, linguists, semioticians and researchers in artificial life, information theory and communication technology.

More information about this series at http://www.springer.com/series/7710

Felice Cimatti

A Biosemiotic Ontology

The Philosophy of Giorgio Prodi

Afterword by Kalevi Kull

 Springer

Felice Cimatti
Dipartimento di Studi Umanistici
Università della Calabria
Arcavacata di Rende CS, Italy

Translated by Fabio Gironi

ISSN 1875-4651   ISSN 1875-466X   (electronic)
Biosemiotics
ISBN 978-3-319-97902-1   ISBN 978-3-319-97903-8   (eBook)
https://doi.org/10.1007/978-3-319-97903-8

Library of Congress Control Number: 2018952473

This Springer imprint is published by the registered company Springer Nature Switzerland AG
The registered company address is: Gewerbestrasse 11, 6330 Cham, Switzerland

Contents

Chapter 1
Introduction

*Life is an incessant imperative for the search of meaning,
something that precedes human reason. Because of this, we
have made the fact of meaning the central problem of
philosophy, capable of erasing any binary division, within the
framework of the evolution of interpretation — that is to say the
evolution of the complexity of systems for reading the world.*

(Prodi 1989: 94–5)

Abstract The natural world, the world of things, is a full world. It is full because everything is in contact with something else, because in the world there is nothing but material events. If the world is such, then what is the *meaning* of a sign? A sign, in fact, is a sending to; it stands for something that is not present. The sign breaks the continuous fullness of the world. Giorgio Prodi tackles this problem, one which is both a philosophical and a biological one by asking how it is possible that, in the material world, something like the *meaning* of a sign becomes manifested. Meaning is not a thing—like a virus or a galaxy—and yet without the notion of "meaning", the biological world would remain incomprehensible. In this introduction, I present the general theoretical framework of Giorgio Prodi's biosemiotic thought.

Keywords Ontology · Sign · Meaning · Semiotics

The world is everything that happens and nothing else. The world is composed by things and events that involve things. A stone, for example, is a thing—something that has a certain place, in a certain time. If the world was merely composed by things, like viruses, we would have no problem in compiling a catalogue of all the things in the world, i.e., its ontology (aside from the decisive fact that, if only viruses

---

This book is a thoroughly rewritten version of *Nel Segno del Cerchio: L'ontologia semiotica di Giorgio Prodi*, originally published by Manifesto Libri (Roma) in 2000. I want to thank Anna Gasperi Campani Prodi for the biographical and bibliographical informations she kindly gave me. Kalevi Kull, Emanuele Fadda, and Carlo Brentari have carefully read a first draft of this book, making it possible for me to improve it in several places.

© Springer Nature Switzerland AG 2018
F. Cimatti, *A Biosemiotic Ontology*, Biosemiotics 18,
https://doi.org/10.1007/978-3-319-97903-8_1

1

existed, there would be no philosophy nor science). Only problems of a practical and technical nature would remain: it is difficult, for example, to describe in detail an astronomical object located many billions of light-years from the Earth. But in the world, there is also something that is not a thing, like a virus. There is *meaning*. For example, where is the meaning of an animal's behaviour? The meaning of a landscape? Of an image? Of the words you are reading? Meaning is not a thing, or at least it is not the same kind of thing that a virus is. But meaning is part of this world. The virus itself, without meaning, could not survive. Meaning belongs in the world, but it is not a thing in the world. This is the problem examined by Prodi: how is it possible to insert *meaning* in the world of things? That is to say, how can we place phenomena related to meaning—semiosic phenomena—within the one world there is, the world of things, the natural world? If the world is full, semiosis digs a hole in it, because semiosis is a sending-off to something which is not present. The sign is here, but its meaning is not: it is *elsewhere*. If we are not willing to renounce the fullness of the world, we have to try and find a way to stitch back together this continuous fabric, lacerated by semiosis. Prodi resolves this apparent paradox by showing how life—biology—is, in the most fundamental of its mechanisms, a meaning, a semiosis. If biology and semiosis coincide, then meaning is not something that punctures the continuous fabric of the world. On the contrary, this very fabric is intrinsically semiosic.

Around the time that Giorgio Prodi wrote *Le basi materiali della significazione* (1977), Umberto Eco wrote that "one must undoubtedly exclude from semiotic consideration neuro-physiological and genetic phenomena" (Eco 1976: 21). For Eco, then, the sphere of elementary biological phenomena is well separated from that of semiosic ones. There is a "lower threshold" (Rodríguez and Kull 2017) of genuine semiosis, a discipline that deals merely with that which is a sign, that is, "everything that, thanks to a previously agreed-upon social convention, can be interpreted as SOMETHING THAT STANDS IN THE PLACE OF SOMETHING ELSE" (Eco 1976: 16). Alongside this definition, Eco proposes another, even more restrictive, one: "semiotics, in principle, is the discipline that studies *all that which can be used to lie*" (Eco 1976: 7). This second principle is much more stringent, since it seemingly excludes from the sphere of semiosic phenomena those, for example, that take place within the immune system of an organism or between an antigen and an antibody (a phenomenon to which the criterion of conventionality cannot be applied). In order to understand the overall philosophical project of Giorgio Prodi, we need to remember how this was meant to challenge and to enter in dialogue with *this kind* of semiotics, developed by his friend and University of Bologna colleague, Umberto Eco. Maturing in this theoretical environment, Prodi is interested, from the beginning, with that which lies beyond the threshold of semiotic: "it is not possible to establish, *a priori*, a semiotic threshold. The field must be completely open towards the origins, and always remain indeterminate" (Prodi 1977: 12). The field covered by Prodi's inquiry will be, on the one hand, that of the natural-biological origins of meaning and of semiosis and, on the other, that of the transformations that meaning undergoes when it becomes a thoroughly cultural and artificial phenomenon. This is the field today known as biosemiotics (Hoffmeyer 1996; Kull et al. 2009; Emmeche and Kull 2011).

Prodi, as a biosemiotician, is also a philosopher: he believes that "semiotics [...] deeply coincides with philosophy" (Prodi 1986: 124). Going from things to meaning—from nature to culture—there is a gradual change, with no abrupt interruptions. It is for this reason that Prodi wants to avoid the two dangerous and opposite pitfalls that beset—today as much as in Prodi's time—the problem of the naturalization of meaning (let us not forget that Prodi's semiotic work is entirely concentrated between the mid-1970s and the mid-1980s. We are here reconstructing the philosophical profile of a scientist, not an episode in the history of semiotic): eliminative materialism, on the one hand, and artificialist culturalism on the other. These are two extremes, useful to better grasp Prodi's peculiar placement.

Let us consider the first pitfall, eliminative materialism, and in particular the stance that considers mental phenomena—that of meaning being its most paradigmatic example—as non-existent. According to a preeminent cognitive scientist, "our commonsense conception of psychological phenomena constitutes a radically false theory, a theory so fundamentally defective that both the principles and the ontology of that theory will eventually be displaced, rather than smoothly reduced, by completed neuroscience" (Churchland 1981: 67). According to this position, meaning would simply not exist—neither in the mind nor in the world. The problem with this kind of materialism is that, in order to be thoroughly eliminative, it becomes unable to account fully for the complexity that it purports to describe (Baker 1989). More precisely this materialism, while attempting to get rid of the notion of meaning, underestimates the absolutely central role of language in human experience, a role that Prodi, as a biologist, is able to fully appreciate. Besides, if meaning does not exist, *who* writes books in order to demonstrate its own non-existence?

The other pitfall is that of artificialist culturalism (a kind of modern version of idealism), a stance that, on the contrary, considers meaning—as, for example, that involved in human cognitive processes—as something completely separate from the natural history of the physical systems that serve as its material basis (like the brain and the human body). As a limit case of this kind of approach, we can look at the work of Nobel Prize winning neuroscientist John Eccles, according to whom "a nonmaterial mental event, such as an intention to move, can influence the subtle probabilistic operations of synaptic boutons" (Eccles 1994: 55). It is evident that a "nonmaterial mental event" cannot be said to belong to the natural world, to the world of material things. If eliminative materialism cannot explain how from things we can reach meaning, a position like Eccles' (of which we should at least acknowlededge its clarity and intellectual honesty) cannot at all explain how meaning is developed from material things—on the one hand, things without meaning and, on the other, meaning without things. Prodi's stance wants to avoid both of these pitfalls. For this reason, his proposal remains timely, since the temptation to do away with meaning is always present, as is the temptation to salvage meaning at nature's expense. As Tim Ingold writes: "the source of the problem is not the conflation of the cultural with the biological, but the reduction of the biological to the genetic"

(Ingold 2006: 276). Prodi works precisely against this dangerous reduction, just as he wants to ward against the opposite risk, run by those who believe that genetics has nothing to do with culture:

> [t]he separation between the biological and what gets called the "spiritual" [...] can be interpreted in two ways. The spiritual could be thought of as too complex to be explained in the vocabulary of the biological, and the biological too rough to be capable of explaining that which is spiritual. [...] These are, clearly, two formulations of the same proposition. One emphasizes the beauty and the perfection of the spiritual — its non-naturality. The other emphasizes the mechanical character of biology. [...] We have preferred a different path, one already looking for some kind of intelligence (nonhuman or anthropomorphic) in the biological, and considering every complication — including logic and rational discourse — as a complication of this intelligence. We called this stance "natural rationalism", identifying it with the elementary semiotics that lies at the foundation of every biological organization. (Prodi 1989: 94)

Prodi's challenge, then, is that of merging continuity and discontinuity, unity and difference, and nature and culture. Prodi looks for a different way, grounded in biology and semiotics, to avoid both materialist monism (there are only things; meaning does not exist) and the dualism of those who decouple meaning from the natural world (there are things and there is meaning, but nothing bridges the two). Perhaps for this very reason, the consequences of Prodi's proposal have not, so far, been explored. That is because Prodi's position is unsatisfactory for both the eliminative materialist and the irreducible culturalist, but most of all because it subverts our unreflective patterns of thought. Let us mention but a few of these theoretical stereotypes: to talk of biology means to negate any historical dimension; the historical-social sphere does not have anything in common with the natural one; human language is a social construction and therefore arbitrary; language is an instrument of communication; the sign is an arbitrary and intentional entity; scientific activity is distinct from poetic production; it is impossible to translate in biological terms the discourse of religious experience; science and philosophy have nothing in common. These, and many more like them, are the stereotypes debunked by Prodi.

Chapter 2
Life

The central moment of consciousness is precisely its theoretical aspect, which is, in itself, praxis.

(Prodi 1974: 197)

Abstract In this chapter, we will present the main elements of Giorgio Prodi's life, and we will reconstruct the entire and complex picture of Prodi as an intellectual: a scientist, a philosopher, a novelist.

Keywords Giorgio Prodi · Scientist · Philosopher · Writer

Giorgio Prodi was born in Scandiano (near Reggio Emilia, in Italy) on the 12th of August 1928, son of Mario, a civil engineer, and Enrichetta Franzoni, an elementary school teacher. The family had humble farming origins, with Mario being the first to get any proper education. Giorgio's eclectic knowledge, his intellectual freedom, and his wide-ranging readings will exercise a great influence on his younger siblings, Fosca (a mathematician), Paolo (a historian of modernity), Quintilio (an architect), Vittorio (a physicist), Romano (the most well-known of the siblings, elected twice as Italian prime minister and president of the European Community between 1999 and 2004), and Franco (an atmospheric physicist). In 1934, the Prodi family moved to Reggio Emilia, where Mario got a job with the provincial administration. Giorgio studied at the *Liceo* Ludovico Ariosto and then enrolled in the Faculty of Medicine and Surgery at Modena University. In his fifth year of studies, he transferred to the University of Bologna where, in 1952, he got his degree in medicine. After that, from 1953 to 1961, he served as *assistente straordinario* of general pathology under Giovanni Favilli. From 1959 to 1961, he worked in Paris, at the *Institut du Radium* (today part of the *Institut Curie*). After having again served as assistant of general pathology from 1956 to 1966, he was eventually called to the second *cattedra* of general pathology at the University of Bologna in 1966, and in 1968, he became Italy's first *professore straordinario* of experimental oncology. He held this position from 1971 to 1978, and in this period (in 1972 to be precise), he

© Springer Nature Switzerland AG 2018
F. Cimatti, *A Biosemiotic Ontology*, Biosemiotics 18,
https://doi.org/10.1007/978-3-319-97903-8_2

created the first institute of oncological research in an Italian university. From November 1978, he was *professore ordinario* of oncology, and in 1984, he obtained a *Laurea* in chemistry at the University of Bologna. There he also created, and served as a director of, an Interdepartmental Center of Cancer Research. Prodi got married twice: first with Anna Maria Nigro—with whom he had a son, Claudio—and then with Anna Gasperi Campani, with whom he had another son, Enrico Emanuele.

Prodi was an intellectual with an extremely wide-ranging cultural interests, who wrote scientific, philosophical, and literary works. He represents, as Umberto Eco celebrated him after his death, a "challenge to the myth of the two cultures" (Eco 1989: 166)—a particularly significant challenge in the Italian cultural context, where this dualism is still strong and persistent. Indeed, Prodi is one of very few figures whose intellectual work science and philosophical inspiration coexisted—nonetheless within an overarching acknowledgement of the humanistic dimension of knowledge (according to the testimony of relatives, and of Eco himself, Prodi would "dedicate a few hours every day to a relaxed listening of classical music" [Eco 1989: 166]). The point of convergence of Prodi's scientific and philosophical-literary activity is probably to be located in his notion of "knowledge" (although Prodi himself insisted that his three "interests"—oncology, philosophy, and narrative—are "pursued separately, paying great care not to merge them together" (Chieco 2011: 611). As Prodi put it, "I believe that science is first of all a theoretical activity that gives us more knowledge about the world" (De Nigris 1981: 27). For Prodi, the principal objective of scientific research is not practical action: science is first of all theoretical knowledge. From this point of view, for Prodi, there is no clear-cut separation between science and philosophy, nor between science and literature. For this reason, Prodi believes that science should be distinguished from its potential technological outcomes: "it seems to me that today", he bemoans, "one of the most negative aspects of modern culture is the belief that science can only be measured through its results" (De Nigris 1981: 17).

Only when disentangled from technology can science be brought back in contact with philosophy and literature. The scientist—like the philosopher and the novelist—does not produce things or goods, but ways of knowing and seeing the world, which can then be turned into actions or practical projects. For this reason, Prodi considered his literary and philosophical activity to be on the same level as his scientific work: he moved seamlessly between his role as an oncologist, as a philosopher, as a chemist, and as a novelist, convinced that there is no single way of knowing the world. With Prodi the overcoming of the "two-cultures" split is not achieved through the demotion of philosophical and literary knowledge to the rank of pseudo-knowledge. When asked to define his profession, he would reply: "I am an oncologist, a researcher", but, immediately after, he would add, regarding literature, "let's say that, for me, it is not just a hobby" (Donati 1985: 16). It is within this general frame that we should place his contributions to biosemiotics,[1] in itself

---

[1] According to Sebeok, the contribution of Prodi to the development of biosemiotic "did not just happen in a simple linear progression but surged by fits and starts as a convoluted affair, winding

something of an oxymoron, at least according to traditional scientific distinctions. Biosemiotics studies biological phenomena *qua* semiosic phenomena, that is to say it merges two spheres that have long been considered separate, the life sciences and the study of culture (of language and of semiosis): Prodi's entire work moves in this direction, attempting to overcome disciplinary oppositions and demarcations. So, Prodi remains a peculiar figure who, precisely because of the multiplicity of his interests, puzzled—and still puzzles—his readers. It is no coincidence that his work has been so well received by Thomas Sebeok, another eclectic thinker with varied interests. Sebeok thus recalls an encounter with Thure von Uexküll, in Freiburg, accompanied by Prodi:

> Prodi, a distinguished oncologist by profession, a novelist, and a prolific contributor to general bio- and endosemiotics — he favored the comprehensive expression "natural semiotics" — had forged, without explicit reference to any other previous or contemporary thinker, still another variant of this sprouting, or re-emerging domain. Prodi was another remarkably creative individual. While the three of us were together in Freiburg (with Thure's sister, Dana, home from Finland, "keeping house"), we conducted an intensive week-long open-ended seminar, so to speak, on the practical and conceivable ins and outs of biosemiotics [...] Our intensive triadic "brainstorming" led directly to the series of pivotal seminars held annually in the late 1980s and early 1990s in Glottertal, on the outskirts of Freiburg. These thought-provoking international get-togethers were held at the Glotterbad Clinic for Rehabilitative Medicine, under Thure's overall aegis and the superintendence of a student and associate of his, Jorg M. Herrman, M.D., its Director. They were attended by many German, Swiss, and other physicians, and were on occasion attended by the biologists Jesper Hoffmeyer (of Denmark), and Kull, now two of the leading figures of the biosemiotics movement. (Sebeok 1998: 34–5)

As I have adumbrated in my introduction, the other pillar of Prodi's intellectual work—beside that of "knowledge" (a knowledge that is also action)—was that of "continuity", an idea to which Prodi gives a very original interpretation. On the one hand, each natural phenomenon is continuous with (and therefore dependent upon) those that preceded it; but, on the other, that same phenomenon modifies the presuppositions that made its development possible—the effect retroactively acts upon its cause. The "continuity" at play here is a to and fro: from the before to the after and from the after to the before. This represents an extremely original element of Prodi's scientific and philosophical work, since it undermines univocal and unilateral notions of causation in the realm of living organisms. This implies, for example, that the division between structure and function, according to Prodi, is biologically meaningless. Take the example of language, a natural phenomenon that Prodi never tired of analysing: "man adapted to language, but he also constructed language [...] it is a snake eating its own tail" (Prodi 1987a, b: 47). Prodi's radical anti-dualism (that however never slides, as often occurs, towards a materialist monism) leads him to privilege the relation over the *relata*. It is not a matter of privileging structure over

---

its long but episodic way through at least three successive twentieth-century iterations: I register these, respectively, with the names of J. von Uexküll [1864–1944], Heini Hediger [1908–1992], and Giorgio Prodi [1928–1987]" (Sebeok 2001: 63).

function, nor the other way around. Meaning—at once biological and semiosic[2]—is that which mediates between the organism and its environment, between the subject and the object, between structure and function. Prodi's philosophy is developed from these two grounding ideas: knowledge and continuity.

In his brief but extremely intense philosophical activity, Prodi developed a genuine "system": he first wrote a general theory of science (*La scienza, il potere, la critica*, 1974) and then a theory of semiosis (*Le basi materiali della significazione*, 1977). He subsequently elaborated a theory—at once phylogenetic and ontogenetic—of the development of various ways of knowing and thinking (*Storia naturale della logica*, 1982), an aesthetics (*L'uso estetico del linguaggio*, 1983a), and an ethics (*Alle radici del comportamento morale*, 1987a). As he observed: "these five books are really five chapters of a unitary argument" (Prodi 1986: 122). The last of this series of books (*Gli artifici della ragione*, 1987b) is an encyclopaedia of sorts, recapitulating in an organic and articulated manner the whole of his philosophical work. His literary works can also be placed in this general framework: *Il neutrone Borghese* (a collection of short stories, 1980) is an ironic and disenchanted description of the omnipresent dangers of fundamentalism, a deadly pitfall for any kind of free and original thought. In the short stories collected in *Cane di Pavlov*, Prodi instead explores that state—between oneiric and allegorical—that precedes knowledge and makes its development possible. The short story from which the collection takes its name assumes the point of view of the dog, overturning the standard perspective: it is not Pavlov who controls the dog's behaviour but rather it's the dog who manipulates and guides him, with intelligence and patience. Here we can see the conjunction of science and literature: the scientist, like the novelist, is he or she who makes us see the world in a new and unexpected way. *Lazzaro* (1985) is a *Bildungsroman*, inspired to the figure of Lazzaro Spallanzani, the scientist and humanist who was born in a palace not far from the Prodi's family house. It is based on one of Prodi's favourite questions: how are things formed? How are they born, transformed, and how do they die? *Lazzaro* is the story of a man who matures a passion for knowledge and science and—as it was observed by Elvio Guagnini (2009)—it is also an autobiography of sorts. Playing a central role in the posthumously published collections of short stories *Dopo il mar rosso* (1990, with illustrations by Cécile Muhlstein) and *Le quattro fasi del giorno* (1987), and particularly in the main story from the former collection, there is the pursuit of something that would come after—as unthinkable and unimaginable that might be—the process of knowledge (that which in the *Uso estetico del linguaggio* Prodi calls, as we will see, the space of "darkness"; *buio* in Italian). In his very last literary work, *Il profeta* (posthumously published in 1992), the lives of two characters are intertwined: a fake Jesus (the prophet of the title) and the oncologist Trequattordici (an explicitly autobiographical character)—the former uses God for his own purposes, his actions having nothing sacred about them; the latter "is religious in every aspect of his

---

[2] In this book I will observe this terminological distinction: "semiotic", as an adjective, refers to the discipline of Semiotics. On the other hand, the adjective "semiosic" refers to the phenomena instantiating a semiosis or a process of signification.

being, but he is without God (and this is no minor absence)" as Prodi writes in the novel. This last claim summarizes the meaning of his entire work, characterized by a passionate epistemological and ethical engagement—all the more serious the less it enjoys a transcendent warranty—accompanied by an implicit "philosophical" conception of life which Prodi, as a scientist and a philosopher, sees as an unstoppable process of translation, construction of new complexes, and continuous transformation (all his narrative works have been republished in 2009 as *L'opera narrativa di Giorgio Prodi*; see Longo 2011). Prodi died of cancer on the 4th of December, 1987, in Bologna. In 1989 he was posthumously awarded a golden medal for scientific merits from the AIRC (the Italian Association for Cancer Research).

Among Prodi's numerous medical-scientific publications, we should mention *Trattato di patologia generale*, written with his mentor G. Favilli (1958 and 1977), *La biologia dei tumori* (1977), and finally *Oncologia generale* (1985). He also published over 200 articles in scientific journals. Among his main research topics, there are the mechanisms of chemical carcinogenesis and of mutagenesis ("In vivo interaction of urethane with nucleic acids and proteins", with P. Rocchi and S. Grilli, in *Cancer Research*, 1970, 30(12): 2887–2892), the role of the response of cell-mediated immunity in tumours ("Selective thymus-derived cell enrichment in the rat spleen as a result of immunodepression by urethane", with A. Di Marco, C. Franceschi, *Cancer Research*, 1972, 32(7): 1569–1573), and the molecular mechanisms of metastasis ("Clones with different metastatic capacity and variant selection during metastasis: a problematic relationship", with P. Nanni, C. De Giovanni, P.L. Lollini, G. Nicoletti, *Journal of the National Cancer Institute*, 1986, 76(1): 87–93). For an overall introduction to Prodi as an intellectual, see G. Mazzoli and S. Zucal (eds.) *Giorgio Prodi e l'avventura del pensare poliedrico*, a special issue of the magazine *Il Margine* (1989), which contains also a bibliography of Prodi's scientific works. Giorgio Prodi was a renaissance man, a scientist who also worked as a philosopher and as a writer, or a writer who also worked as a scientist and as a philosopher—Prodi displays the same critical engagement in each of his intellectual productions. To this fundamental coherence of his thought, we should add the fact that Giorgio Prodi was unlike other philosophers particularly considering those whom today we consider philosophers: those who, in order to claim mastery of a "scientific method", copy the style of scientific papers and fill their writing with references and citations. Prodi, on the other hand, hardly ever cites another philosopher[3]—yet he was not a thinker working in isolation. For example, in his 1988a essay "Material Bases of Signification", one of the very few of his articles accessible for an international audience, Prodi (as Sebeok observes) only cites Frege and Ogden and Richards' *The Meaning of Meaning*. Although this, of course, does

---

[3] "Another eccentricity of Prodi is his avoidance of references to the works of others. For example, in his English article, although dealing with intrinsically biosemiotic issues, viz. of "natural semiology" (1988a: 206), he cites only Frege and the 1923 edition of Ogden and Richards. While this composition style perhaps adumbrates Prodi's striking originality, it fails to align him with any predecessors or successors in semiotics, so his untutored readers may flounder for lack of familiar signposts" (Sebeok 2001: 68).

not mean that his semiotic theory is solely inspired by these authors, Frege is the logician who first distinguished between "sense" and "meaning" (or "reference"), a distinction that is systematized in the famous semiotic triangle (Ogden and Richards 1923: 11). Without the semiotic triangle, there is no meaning, and without meaning, there is no semiosis but only causal connections: this is why these two sources are essential for any project attempting to develop a theory of biosemiotics. However, Prodi generally follows *his own* path, precisely because he doesn't feel the need to demonstrate that he is a philosopher—he simply *is* a philosopher—"Giorgio Prodi [...] was a maverick: a prolific physician and experimental oncologist by profession, a novelist by avocation, but also an intermittent if resolute contributor to biosemiotics" (Sebeok 2001: 67). In a relevant passage from his Preface to *La scienza, il potere, la critica,* Prodi writes that:

> the standard way to construct an argument is now "by proxy of a bibliography". Every discourse becomes an academic game of chess, a system of oblique cultural references. The discourse in this book wants to be more direct. Practitioners of philosophy will recognize the general frame, the major positive references (Darwin, Freud, Marx, Dewey) and the main negative ones (Hegel, Husserl, Bergson), as well as the closest and more concrete sources of inspiration (Piaget, Russell, Wittgenstein, Morris, Carnap, and Chomsky, to cite but a few). They will also recognise various -isms: empiricism, structuralism, idealism, irrationalism, and so on. Those who are not philosophically proficient will follow my discourse with greater ease, not bound by the intellectual complacency of being able to recognize the characters behind the text. (Prodi 1974: 6)

What was Prodi's legacy, in Italy as well as abroad? It is necessary to consider that he worked during a period of profound changes in the Italian cultural and political landscape and approached themes—like that of the natural basis of knowledge— that were extraneous to the dominant philosophical trends of the time: historicism, Marxism, structuralism, and phenomenology. Moreover, Prodi was not a "specialist", not in philosophy nor in semiotics, and those who are able to move between different disciplinary fields are viewed—then as today and not only in Italy—with suspicion. This diffidence has grown with time, especially so with the progressive specialization and fragmentation of the human sciences. Contemporary scientists and philosophers are, typically, specialists in *one* disciplinary sector. Although it is customary to pay lip service to "interdisciplinarity", this is never actually appreciated, and even less it is ever put in practice. Prodi was an oncologist, a doctor-scientist, but also a philosopher while also being a novelist. From this point of view, he was far too different from the average professional philosopher or scientist, and it is all too natural that he was met, then and now, with suspicion or indifference.[4] Prodi was well aware of this. As he comments in a 1986 interview:

---

[4]One of the few exceptions to this silence regarding Prodi's work is Pierpaolo Antonello's fine book *Contro il materialismo. Le "due culture" in Italia: bilancio di un secolo* (2012, see in particular pp. 206–8 and 302), reconstructing the history of "materialism" in the twentieth-century Italian culture. Antonello associates Prodi to another rather forgotten figure of Italian post-war philosophy, Ferruccio Rossi Landi (1992). For Antonello, Rossi Landi and Prodi proposed a "materialistic semiotics" that was neglected during their lifetimes. Indeed, Prodi's case is exemplary still today: his "sophisticated" materialism, as I adumbrated above, has little in common

[my monograph *Le basi materiali della significazione*] has been read with interest by a number of attentive semiologists, and well received by a few epistemologists but, overall, the Italian philosophical environment — to which the book was explicitly addressed — was, and still is, profoundly resistant to these kinds of problems. [...] Why? I think there's primarily two reasons. First, disciplinary corporativism (not only by philosophers, of course): it seems that, in our context, to pay attention to a philosophy book written by a non-philosopher would be a reckless enterprise. Secondly, our theorists tend to busy themselves only with "authorized" — so to speak — topics, possibly non-philosophical but para-philosophical (the death of philosophy, negative thought, weak thought, strong thought, half-and-half thought, the sociology of intellectuals within their institutions, and so on). At times, when I think about philosophy, influenced by those youthful suggestions one always remains faithful to, it appears to me as an old lady, of noble linage but now in dire straits and rather hungry, forsaken by her own sons. There is nothing wrong if distant relatives (perhaps biologists) want to take care of her, keep her alive and pay their respects. (Prodi 1986: 123)

1. China 1956

with, for example, the implicit or explicit materialism of contemporary neuroscience which, criticizing Cartesian dualism, gets rid of all semiosic phenomena and of meaning in general. Prodi was an untimely thinker for the philosophies of the subject and of culture of the second half of the twentieth century. For other, rare exceptions, see Caputo (1990) and Zorzella and Cappi (2012).

2. 20th July 1985

3. At a oncological congress in Bari, June 1980

4. While receiving the Sila Prize, 1986

5. Prodi giving a lecture on "Philosophy of Knowledge and Biology", University of Udine 1983

Chapter 3
Scientist Because Philosopher, Philosopher Because Scientist

> *Critical thought does not begin with a suspension of the usual modalities of knowledge, thus inaugurating a new, purely human, course. Rather, it reflects more carefully on those modalities as they relate to things, therefore grounding itself in the real. Science and critical thought are therefore synonymous.*

> (Prodi 1974: 147)

Abstract Why did Prodi, while engaged in his scientific work, begin to study and to write philosophy? In this chapter, I will attempt to answer this question. The guiding idea is that Prodi realized that, in order to truly be a scientist, one needs to reflect on what it means to be a scientist. Otherwise, the scientist loses the intrinsically ethico-political dimension of his or her work. But once Prodi stepped into the philosophical field, he could not stop, since—consistent with his belief in continuity—he realized that every philosophical problem is intimately connected with others. Prodi the scientist becomes Prodi the philosopher.

Keywords Science · Philosophy · Critical thought · Darwin

Giorgio Prodi was a scientist—an oncologist—and precisely in his role as a scientist, he began to be preoccupied with the scope of his scientific practice: questioning what he was doing when designing an experiment, how to conduct it, and what discoveries it would allow (when successful), that is, what the nature of the object of the experiment was. But his main interest was answering the question: what makes knowledge possible? These are not the sort of questions usually asked by the average scientist, too busy with the immediacy of laboratory work, with observation, and with the analysis of experimental results. But Prodi must have been a curious kind of scientist, in the philosophical sense of the word—that is, a person that is not content with doing something but who also asks why he or she is doing it. His questions are philosophical, that is, questions about the meaning of a certain activity, rather than about how to perform it, or about its proximate outcomes. These are

© Springer Nature Switzerland AG 2018
F. Cimatti, *A Biosemiotic Ontology*, Biosemiotics 18,
https://doi.org/10.1007/978-3-319-97903-8_3

the kind of questions addressed by many philosophers, but their answers likely did not satisfy Prodi, since his new philosophical adventure began precisely from the attempt to formulate new answers: in just a few years (from 1974 to his death), he developed a complete—and rather original—theory of knowledge and of biological reality as a whole. A theory that reinterprets in a biological key semiotics and the theory of knowledge, and through them ontology as a whole.[1] In this chapter, before attempting to present at least some of the ideas contained in such a vast and complex body of work, I want to try to explain why Prodi found the common philosophers' answers so unsatisfactory. It is common to believe that the problem of knowledge pertains, roughly put, to the encounter of the knower—the subject, according to traditional terminology—with that which he or she is experiencing, the so-called object. How can the former know the latter? That is to say, the problem of knowledge would be that of reconciling two poles of a fundamental dualism, that between the subject and the object. Such a radical dualism is irresolvable, since no matter how much effort on our part, the object will forever remain too distant from the subject: the subject is separated from the object because of its different material composition (as in Eccels' case, as we just saw. See Lavazzo and Robinson 2014), because it is governed by different principles (the first is moved by reasons, the second by causes), or because it acts according to different modalities (the first is active, the second is passive), and so on. In all these cases, one could never offer an explanation of the puzzling fact that knowledge works: that is, it has real effects upon the world. This dualism cannot be recomposed, and knowledge—something that a scientist like Prodi experiences every day—remains inexplicable, as a kind of embarrassing mystery (at least for some philosophers). Prodi, on the other hand, gave an apparently simple but profoundly radical answer to the question of the possibility of knowledge: an answer that, instead of duality and discontinuity, presupposes continuity and the unity of biological systems. For Prodi "knowledge 'emerged' from things, and it can know them since it is shaped by them and shares their same origin" (Prodi 1974: 134).

It is not we, presumptuous subjects, who know the object. Rather, things let themselves be known by us, subjects who are nothing but the ultimate transformation of other things, in turn connected with more things, and so on all the way to the very objects we are knowing. The world does not include an a priori distinction between subjects and objects, which would be separate and autonomous, but only one between more or less complex organisms, linked by an infinitely articulated fabric of relations, which coincides with life (and semiosis, and therefore knowledge, as we will see). If every dualism is based on the model of two separate and opposite poles, Prodi's model is a biological one, grounded on the evolution of every form of life, wherein no gap ever exists, but only transformations and transla-

---

[1] For this reason, Prodi's theory is an ontology, that is, a description of the structure of the world, based on biosemiotics: a systematic bio-ontology. Prodi's perspective is similar to that of Buchanan (2008), although his book is never cited. According to Buchanan, "ethology emerges as the significant dimension in framing the being and becoming of the animal. The animal body is interrelated with its environment through the process of behavior, so it becomes a question of how to engage the ontological dimension of this relation" (Buchanan 2008: 5).

tions of shapes into other, more or less complex, shapes. Subject and object can never constitute the starting point of the process of knowledge. They can at best represent its end state: one never fully realized, since such a complete separation would break the biological continuity that binds them together—and this would entail not just the end of knowledge or of semiosis but of life itself. The dualist model can be schematized as based upon two distinct points that need to be somehow connected; the biological model, on the other hand, can be represented as a circle (or a spiral; see Chap. 10), since there is no life form that does not derive from another life form and that does not ultimately return to life itself. Knowledge, then, is not a view from the outside towards the world, but it is the gaze that each particular thing—the thing which the philosophers call the "subject"—casts upon some other thing, the so-called "object". Knowledge is a relation between worldly things.

Modern philosophy—starting at least with Descartes—seems to have been dominated by the model of the straight line and of the direct opposition, and to exhibit a certain fear of the circle, often accused of being a "vicious" one, because it always returns to its starting point. Prodi's philosophical work can be interpreted as a systematic application of the model of the circle to semiosic and biological phenomena. Where the straight line presupposes and reproduces both a separation and a distance between two points located at its extremities (subject and object, mind and body, spirit and matter), the circle has no beginning and no end. Life indeed has, strictly speaking, no beginning, since it is best described as a continuous process of transformation: from forms to different forms. Life does not begin in a body, at moment *t*, as if at moment *t-1* there was still no life. Schrödinger's definition illustrates this point, explaining how, and in what circumstances, a random "piece of matter" can be considered "alive":

> [w]hat is the characteristic feature of life? When is a piece of matter said to be alive? When it goes on 'doing something', moving, exchanging material with its environment, and so forth, and that for a much longer period than we would expect an inanimate piece of matter to 'keep going' under similar circumstances. When a system that is not alive is isolated or placed in a uniform environment, all motion usually comes to a standstill very soon as a result of various kinds of friction; differences of electric or chemical potential are equalized, substances which tend to form a chemical compound do so, temperature becomes uniform by heat conduction. After that the whole system fades away into a dead, inert lump of matter. A permanent state is reached, in which no observable events occur. The physicist calls this the state of thermodynamical equilibrium, or of "maximum entropy". (Schrödinger 2013: 69)

According to this physical picture, there is no special essence, proper and exclusive of "life", which mere matter would not possess. A phenomenon is an instance of life if it remains itself and changes itself *at the same time*. There is no life if that "piece of matter" becomes incapable of transforming itself, while retaining its identity. For this reason, the model of the circle is well-suited to describe life phenomena, since the circle does not proceed towards a direction, only to stop once the destination has been reached. The circle always repeats its perpetual movement: the circle *is* nothing but movement. Prodi's predilection for the circle, and rejection of the fixity of the straight-line model, probably derives from a book by another important Italian philosopher, Enzo Melandri's *La linea e il circolo. Studio logico-filosofico sull'analogia*, first published in 1968. The circle is usually criticized for being "vicious", always going back to its starting point. But is this accusation justified? The straight line is the model of dualism, while the circle is the model of analogy and therefore of mediation and of continuity. Why would opposition and dualism be more relevant—from a logical and an ontological point of view—than the principle of analogy, that is, the idea that the connections between things (thoughts or objects) are always gradual, blurry, and continuous? Why assume that the separation is at the origin of the process of knowledge, rather than knowing relation itself? The two poles—subject and object—which are the endpoint of any dualism "have this problem: by establishing a radical dichotomy they make it impossible to comprehend the matrix from which the terms of the relationship depend. This comprehension requires instead that the point of origin be given precisely by that which appears — when starting with two opposite poles — like a *tertium comparationis*: the principle of analogy" (Melandri 2004: 792). This *tertium* would really be more than a simple analogy: a generative space upon which the internal articulation of the other two poles depends. But to rehabilitate the principle of analogy—a sort of logical equivalent of the principle of continuity in the biological domain (later we will see how continuity does not imply gradual evolution; see Chap. 7)—means precisely to favour the model of the circle over that of the straight line: "the metaphor of circularity should be given a more reckless interpretation. The criticism of the circle as a 'fallacy' in itself represents a certain 'rectifying' fallacy, with remote metaphysical origins. Why would a linear order be preferred to a circular one? Why the line and not the circle?" (Melandri 2004: 794).

The image of the circle also helps understanding Prodi's peculiar way of being a philosopher and a scientist and therefore a scientist and a philosopher. In fact, this is the most challenging facet of Prodi's intellectual outline—but also the most important one—a difficulty that explains why his theoretical work has been so easily forgotten. The crucial point is that, for Prodi, it is impossible to be scientists without also being philosophers and vice versa. Today's science is thought to be a special activity and philosophy to be a completely different endeavour, a purely speculative—and therefore inferior—activity (it is no coincidence that older scientists, too old to be at the cutting edge of their discipline, attempt to write philosophical books, while the opposite does not happen: an elderly philosopher does not fashion himself or herself as a scientist). Science today has gained such an elevated position of autonomy and veneration that any other kind of discourse is considered inferior. Prodi shows us how wrongheaded this is, because a scientist who is not also a philosopher can never be a good scientist. The premise of this stance is explicitly stated in the opening pages of his epistemological treatise: "what is common to all those discourses that define science as a practical endeavour, aimed at transforming the world, is their lack of understanding of what science is" (Prodi 1974: 5).

In particular, the scientist (but also the philosopher) does not understand how his or her activity is not isolated from, on the one hand, the rest of human epistemic activities and, on the other, from the biological origins of human knowledge. In sum, Prodi takes Darwin's gradualist approach seriously (Mayr 1982: 508–9): the scientist, like the philosopher, is nothing but one of the many, infinitely varied, forms assumed by life on earth. The scientist, from this point of view, is nothing but a mode of being of a single underlying vital principle:

> there is grandeur in this view of life, with its several powers, having been originally breathed by the Creator into a few forms or into one; and that, whilst this planet has gone cycling on according to the fixed law of gravity, from so simple a beginning endless forms most beautiful and most wonderful have been, and are being evolved. (Darwin 2006: 429)

Darwin's "creator" is life itself, that is to say this continuous process of transformation. It is for this reason that the scientist is also a philosopher, because science—its prestige and effectiveness notwithstanding—is still a human, animal, biological activity: in the final analysis, a vital activity. Prodi as a philosopher is fully conscious of this absolute biological continuity. Before the appearance of the scientist and the philosopher, there is the world of life, one composed by change and relations. Prodi, as a philosopher, understands that life means relation, contact, merging: "the starting point [...] is therefore nothing but the recognition of a structure-thing nexus, that is to say the existence of a system of interactions. [...] In other words, we should start by assuming the physical-biological character of the epistemic process" (Prodi 1974: 27). But what does this concretely mean? It means that the knower, the subject, is always already involved in that which is known: that is, the subject is nothing but a continuation, in a different form, of the object. Prodi writes: "if the object is 'known' as involved in a network of relations and part of this structure—and therefore never as an isolated and constant element—then it will be impossible for it to be seen 'at a distance' or to be 'criticized'. Rather, it can only be immediately manipulated and consumed" (Prodi 1974: 43). It is in this sense that the subject is *taken* by the object and the object is *absorbed* by the structure of the subject. This model foregrounds the *relation* (at once epistemological and biological), by deprioritizing the *terms* of the relation.

By foregrounding, in his philosophical thought, the concept of relation—as a vital field wherein the organism meets its environment and where the latter envelops the former—Prodi ends up questioning his own work as a scientist and particularly as an oncologist. The starting point is always the same: in nature, no entities or activities are completely separate and autonomous. Knowledge, including its most specialized and sectorial forms, is part of a larger ensemble of various types of epistemic "metabolisms", proper of different living organisms. The moment the scientist becomes aware of this radical non-separation, then he or she steps into philosophical terrain:

> [K]nowledge is a biological process of interaction, and its methods are relative to things and to structures. [...] If this field of action is circumscribed as a scientific object — that is, if knowledge is applied to things with an eye towards them and one self-consciously toward its ways of check [riscontro] — it then becomes critical thought: a self-examination in relation to other things, a self-analysis, and self-critique. Critical thought is not a secondary stage of the epistemic process, or a kind of super-knowledge. Rather, it is an additional differentiation of a natural evolution, reaching the point of turning the entire system and structure of things into an object of knowledge, analysing its reciprocal interactions, the characteristics and limits of the check, and its principal internal modalities. (Prodi 1974: 146)

Philosophy then would not be a supplementary, additional, and merely theoretical form of knowledge, subordinated to science. For Prodi the philosopher is a scientist who has thematized his or her own role as a scientist. The dualism between science and philosophy is therefore finally decommissioned. Prodi's monograph *La scienza, il potere, la critica* contains, in an embryonic and implicit form, the entirety of Prodi's later philosophical output. For example, once the scientist poses the question "what am I doing, when practising science?", he or she is clearly stepping into the proper domain of ethics. The problem of the scientist's responsibility—of ethical or ecological nature—is not an additional issue, supplemental to his or her more straightforwardly experimental scientific work. The scientist-philosopher cannot but formulate this question, for otherwise he or she would not be a proper scientist, since "morality is nothing but an aspect of knowledge" (Prodi 1979: 187). The scientific attitude is intrinsically ethical. More specifically, as we will see later on (see Chap. 8), the objectivizing gaze of the scientist presupposes the ability to take care of the object studied, to isolate it from the network of relations it is embedded into, and to treat it as something special. But this already constitutes an *ethical* way to look at the world: "it is knowledge itself that, evolutionarily speaking, needed this attitude in order to become reflexive and propositional—what we call morality. Morality is an intrinsic component of knowledge and of the development of the human mind" (Prodi 1987a: 42). This is like asking what the origin of this attitude towards the world is: hence Prodi's research into biosemiotics, tackling the question that opened this book—how is meaning possible in the world of things? (Sebeok 2001; Barbieri 2009). To know means to attribute a meaning to events. A certain event has a meaning, while some other does not: to know means to distinguish and to set apart phenomena that are relevant for a certain form of life from all the other meaningless ones. A pertinent event for a form of life is a significant event, an event with *meaning*. Prodi then, as scientist-philosopher, asks: what is the natural origin of meaning?

> [t]he elementary case of an enzymic reaction allows us to formulate a more general argument. If we consider the enzyme as a pre-given structure (with a long evolutionary history, even in the most elementary of cases) this entity explores and knows only one section of the world, in a restricted and determinate space, its own substrate. It only reacts to (that is, it knows) what corresponds to that onto which, evolutionarily, it has modelled itself. Its knowledge has an extremely narrow range. Its capacity for exploration and manipulation is rigid and automatic. But what is the metabolic knowledge that the enzyme has of its substrate if not the discovery, among the indifferent things in comes in contact with, of what is complementary to itself, that makes it move, that has a meaning, that is a sign? (Prodi 1974: 231)

Now, if life is tantamount to being in relation to a meaning—i.e. acquiring knowledge—the problem of the *internal* limits of knowledge cannot be avoided. Once again, this question is nothing but a development of the biological premise of Prodi's reasoning. To know means, as we just saw, to distinguish and to attribute a meaning to phenomena. But this also means that every act of knowledge leaves behind a residue: everything that is not meaningful to a form of life in a particular moment of its biological development. Therefore knowledge at the same time produces ignorance.

Prodi calls this residue "mystery" or "darkness" [buio]. This "mystery" is not something irrational and unthinkable; on the contrary it is the product of the internal functioning of reason and of thought. We can know something only because we are within it: only in this way can we distinguish that which—from our point of view—is meaningful from that which is meaningless. But every "within" implies and presupposes a radical "without": "if the world is much larger than our frame, and it dwarfs us physically and historically, then our inclusion within it is the crucial node of both existence and knowledge. There is always a without, a mostly dark area, and it is unthinkable that that could ever be extinguished" (Prodi 1974: 169). Once again Prodi, as a scientist-philosopher, finds it necessary to pose the question of the "mystery", that is to say to question the ultimate meaning of human experience, now no longer exclusive province of philosophers or theologians. Prodi can then establish the problem of the sacred in purely scientific and biological terms, because the scientist who reflects critically on his or her scientific work is, in a certain sense, forced to do so, since the problem of the limits of knowledge is inseparable from the question of knowledge itself. So, the problem of the sacred—like the question of the "mystery" and the "darkness"—is internal to the scientific enterprise itself: "so the mystery is that which is utterly unsayable, potentially open to knowledge, lying on the horizon of our interactions" (Prodi 1974: 170). It needs repeating then that the "mystery" is unsayable not because it is irrational or because it would go beyond our thought and our expressive resources: on the contrary, it is unsayable because it is located in the middle of the sayable, that is, of scientific knowledge. Science continuously produces "mystery". Therefore, there is nothing more risible than those scientists, unaware of their own doing, who attempt to "scientifically" resolve the problem of the "mystery" (like those who endeavour to "prove" or "falsify" God's existence). The truth is that "mystery", like the horizon, recedes the more we get close to it. The horizon is not unreachable because of some unknown force, hindering our progress: we ourselves are the "mystery" of the horizon.

Lastly, there is also a political consequence of this stance. The image of human nature that emerges from Prodi's thought is that of an animal who is in a perennial state of crisis, since the fundamental operation of meaning attribution also produces, time and again, its opposite—that which is meaningless: "essentially, then, our very nature is constituted through this crisis" (Prodi 1974: 384). The human is the living being characterized by a "constitutive crisis". It is therefore an unsettled, dangerous, and curious animal, always faced with the "mystery". As usual, it is necessary to track the biological origins of this living being: if life means meaning and knowledge, the particular way of knowing of the human animal largely depends on language—"the immediate expression of the intersubjective character of knowledge is language. The possibility of communicating, in any way, is part and parcel of the process of knowledge. It is impossible to imagine a scenario in which knowledge would develop as a self-referential relational possibility, and only later would become communicable. Communication and knowledge constitute one and the same process" (Prodi 1974: 223). But "intersubjectivity" means different points of view, clash, and conflict. This is the bio-epistemic motivation for politics: nothing but the attempt—constantly on the verge of failure—to come to terms with such a restless organism. Prodi writes: "since human nature is essentially a social and collective product, there is nothing that can be produced by it that is not political. In this sense, politics is itself a scientific-epistemic activity" (Prodi 1974: 402). At the roots of this reasoning, there is, once again, the human's peculiar natural history: "language and thought belong to human biology [...] [and are its] distinctive traits" (Prodi 1989: 92). But this means, as we have just observed, that human nature is itself a synonym of "crisis" and "mystery". Herein lies the biological origin of politics (and obviously not in the social behaviours of primates). Prodi's political reflections are not an additional, artificial addendum to his scientific and philosophical work: rather, there is politics because in every "enzymatic reaction", there is a choice and therefore meaning and conflict—in short, politics.

Chapter 4
The Line and the Circle

It is initially assumed that knowledge is achieved by (consists in) check [riscontro] operations, where the knower is modified by the surrounding environment. The existence of a network of interacting material situations is then postulated. The object that can receive knowledge-producing feedback, which we call here the "reader", is one of the many elements that compose this network of changes, being enmeshed in it. To assume the network means: "the starting point is the network" or, more generally, "there is a network, and knowledge is part of it".

(Prodi 1982: 15)

Abstract This chapter describes Prodi's peculiar strategy to explain life phenomena. This is a model based on the concept of relation or biological meaning that always privileges the relation over the terms of the relation. For Prodi a relation is never a one-way affair, but, on the contrary, it is always a to and fro. This is because the general epistemic model for life phenomena is that of the circle, a figure that has no preferred direction. This method allows Prodi to avoid any kind of determinism, be it genetic or cultural.

Keywords Natural meaning · Nature · History · Material correspondence

At the most general level, to explain something means to look for a fact or principle which would be simpler than that which needs explaining—a fact or principle which would serve as grounds for the *explanandum*. A good explanation would then possess one essential characteristic: if the *explanandum* is located at the ontological level L, the *explanans* should be found at least at level L-1. That is to say, the explanation needs to appeal to simpler and more basic facts and principles than the *explanandum* does. The most important element, then, is the simplicity (at least in principle) of the explanation. Let us make an example. We have to explain why the males of many species of birds have particularly flamboyant plumage—a physical characteristic that could be a lure to predators—as well as being often large and

© Springer Nature Switzerland AG 2018
F. Cimatti, *A Biosemiotic Ontology*, Biosemiotics 18,
https://doi.org/10.1007/978-3-319-97903-8_4

cumbersome (as in the case of the peacock's tail). A plausible explanation is that this is meant to attract the attention of females and that such a highly visible plumage helps them find a mate. There is a fact F—the male's flamboyant plumage—and a measurable effect, since it can be demonstrated that a female will gaze more intently at a bird with flamboyant feathers than one which does not have such a plumage (Zahavi, Zahavi 1997). If this counts as a plausible explanation of F, that is because it is simpler than F; the explanation E, in fact, does not presuppose complex principles. At its core, the theory of evolution by natural selection states that animals compete among themselves for survival and that those who—in a given environment—are bearers of the physical and behavioural characteristics most suitable to it manage to survive and therefore to reproduce, thus transmitting their "advantageous" genes to successive generations—while others do not. Some might dislike it, but this is a very simple explanation indeed, and a very powerful one too, since it can explain a multitude of different facts. The most interesting aspect of this kind of explanation, for our purposes here, is that in order to be efficacious, it does not require the intervention of the subject whose behaviour is explained by it. That is, it is not necessary to presuppose that male birds have an explicit intention to attract the females with their flamboyant plumage: whether they want it or not, their body is composed in such a way that it will attract the females' attention. A good explanation, therefore, does not employ notions like will or intention. A good explanation is non-intentional. As a matter of fact, in all those cases where a will or an intention comes to play a role in an explanation, this loses its essential character—simplicity. Nothing is harder to comprehend than the notion of "will" or "intention". An explanation based on "will", that is, is more complex than the fact it purports to explain.

When this explanatory logic is applied to mental phenomena—and in particular to semiosic ones—it seems natural to suppose that it would fail to be adequate. It appears obvious that, for this kind of phenomena, an intentional explanation would be necessary: in the sense that F would be explained by presupposing an explicit intention to give rise to F. For example, we may wonder why newborn babies cry: according to an intentional explanation, their cries are caused by an intention in their mind (albeit a confused and indistinct one), that of attracting their parents' attention. Mental phenomena, and in particular semiosic ones, would then only be defined by—and explainable through—the intention of those who have them (Buyssens 1943). This thesis is buttressed by the idea that the employment of signs (at least in line of principle) is based on an explicit—that is to say voluntary—convention. For example, the fact that, in English, the neighing equine is called "horse" depends on a "convention", that is to say on an explicit decision taken by a group of human beings. In general, according to this explanatory strategy, targeting semiosic phenomena (and in particular human ones), the best explanation would be an intentional one.

However, this strategy has a fundamental defect: it does not respect the principle of explanatory simplicity, since the *explanans* is more complex than the *explanandum*. If F is at level L, E—in this case—is at level L + 1. We have just seen why: we do not have any access to that which occurs inside people's heads, since we do not know anything about their intentions. Indeed, we do not even know if intentions really exist. Thus, the *explanans* is more complex than that which it purports to explain and, properly speaking, it is not an explanation. Prodi's fundamental methodological choice, coherently with his predilection for biological continuity, is to avoid this kind of explanation altogether. At the same time, however, Prodi wants to avoid the opposite pitfall: those explanations that simply 'eliminate' the fact that should be explained. He therefore follows a two-tiered commitment: he wants to formulate explanations of complex phenomena, but such explanations must not be reduced to the claim that the complex phenomenon to be explained, in fact, does not exist (as we have seen in Chap. 1, vis-à-vis eliminative materialism).

In order to avoid both of these pitfalls, Prodi always looks for the explanation of a phenomenon in its evolutionary history. This strategy is advantageous because evolutionary history is not a one-way phenomenon: on the contrary, it is a paradigmatic process wherein organism and environment are linked together in a relation of *complementarity*. Here, as will often be the case in the course of this book, we come back to the most simple of situations, when an elementary organism categorizes something in its environment as meaningful, bracketing everything else. As we have seen, this, for Prodi, is an instance of knowledge, indeed the very matrix of all forms of knowledge: "there is a profound relation between the knower and the things he or she can know. This is a relation of derivation, that is to say a genetic one [...] broadly conceived. Every living being 'knows' the world to which it is adapted and from which it derives. To know, in this radical sense, means to interpret the environment, to move within the environment, and to survive in the environment" (Prodi 1979: 182). Complementarity means that between the organism and the thing, a relationship occurs and indeed that the relationship itself fixes the organism as a "reader"[1] (with respect to the thing) and the thing as a *meaningful object* (for that "reader"). So, "to know a thing means to be changed by it. Knowledge is always, at every level, a process through which things change, and are reciprocally adapted" (Prodi 1979: 185). The crucial claim is at the end of this quote: the organism and the thing are "reciprocally adapted" entities. The "reader" is adapted to the thing, and the thing is adapted to the "reader". This is the model of the circle, applied to the world of living beings. Returning to the sphere of human semiosic phenomena, it is now clear why they cannot be explained with intentional explanations:

> It is necessary to reject an interpretation of the term "subjective" which pits it against "objective". The usual anthropomorphic criteria that assume intentionality and consensus as primordial facts, are inadequate. [...] If we assume, as a criterion used to define the field of semiosis, intentionality qua condition of proof (suggesting that semiosis would begin with the "will to communicate", and therefore clearly distinct from other natural functions) we are operating anthropomorphically. The criterion we are using to demarcate the field is "consciousness", with all the ambiguities that burden this term. So, the attitude of the semiologist, rejecting as extra-semiotic all psychological and biological influences (broadly conceived), [...] is conditioned from the start by weighty psychological or even introspective presuppositions. Upon examination, the facts of "consciousness" immediately appear as rooted on natural bases, unconscious and automatic, and can only exist if grounded on these: consciousness is the tip of the iceberg, and if we want to explain anything (in the simple sense of connecting it with something else) we need to refer to the submerged part, to that which allows the tip to emerge. (Prodi 1977: 18)

---

[1] According to Prodi any living entity "reads" the surrounding world, that is, it "selects" the meaningful elements of it in respect to its own interests.

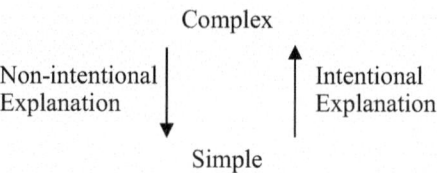

Fig. 4.1 Two kinds of explanation: intentional and non-intentional (This schema, like others that will follow, is not derived from Giorgio Prodi's works. Just as he never explicitly cited his sources, so in his often compressed and dense prose, he never tried to explain too much: "without an effort to deepen knowledge, and its systematic organization into a science, it is impossible to understand the specific character of the times we live in" (Prodi 1974: 7))

Traditionally, semiotics, as we will see in what follows, has attempted to explain semiosic phenomena by means of intentional explanations, that is to say positing at the origin of semiosis a subject that voluntarily choses to use a certain material object—the sign—in order to signify something else. This, has I have argued, is not an explanation since it is based on a more complex notion than the fact it purports to explain: "consciousness is a fact to be explained by means of non-conscious facts, even though it is customary to invert the terms of this problem" (Prodi 1982: 108). Conversely, Prodi tries to explain the whole of natural semiosic phenomena— from the most elementary to the most complex, like human language—by means of non-intentional phenomena, i.e. respecting the constraint that an explanation needs to be simpler than that which it explains. While intentional explanations can be represented as an arrow pointing downwards, from complexity to simplicity, non-intentional explanations can be represented as an upturned arrow, from the simple to the complex (Fig. 4.1).

However, this model is merely a first approximation of Prodi's explanatory style. Here we still see lines and arrows, while we have already detailed how the scientist-philosopher looks for a relation of "complementarity". However, as we will soon see, it takes little effort to transform this model into a circle. It requires the integral adoption of a biological point of view.

Let us go back to semiotics, a discipline that in Prodi's days was mainly concerned with institutionalized systems of signs and of systems, that is, of signs currently in use. It must be remembered that Prodi was working during the final days of the age of structuralism. The core idea of this movement was that what really matters in signs is the *structure* holding them together. The question of the origin of the semiosic link between signifier and signified was considered uninteresting. For the structuralist, the origin is never a scientific problem, as, for example, de Saussure claimed:

> to distinguish between the system and its history, between what it is and what it was, seems very simple at first glance; actually the two things are so closely related that we can scarcely keep them apart. Would we simplify the question by studying the linguistic phenomenon in its earliest stages—if we began, for example, by studying the speech of children? No, for in dealing with speech, it is completely misleading to assume that the problem of early characteristics differs from the problem of permanent characteristics. (de Saussure 2011: 8–9)

However, this approach leaves the fundamental problem of semiosis unexamined: if the sign is essentially a sending to, what guarantees the link between the sign and the object to which it refers? Prodi's answer to this issue—the reply of a scientist-philosopher—is to look for the origins of semiosis not in culture (so in social convention, communicative interaction, and intention to transmit a thought), but in biology, and more precisely in the relations that obtain between different cellular and intracellular entities. The key concept mobilized by Prodi in this venture is that of "reading". Every organism puts itself in a relation with the surrounding environment by selecting the characteristics that, from its point of view, are pertinent, i.e. *meaningful*. Natural meaning is the result of this reading operation. Prodi employs the concept of "reading" or "interpretation" because he wants to highlight the organism's propositive role vis-à-vis the environment. To read means to attribute a meaning; at the same time, the meaning of what which is read cannot be completely arbitrary, because the world affords only some possible readings and not others:

> In nature, meaning arises as a correlation between an organism and a section of the world that can be interpreted thanks to the constitution of adequate structures to interpret them. A thing becomes meaningful when it can be deciphered by someone (that is to say: exploited in a specifically metabolic sense, for survival etc.) Biological objects [...] (*the readers of things*) are modelled by environmental conditions, in the sense that things function for them as a reference point, a filter, the reason for their evolution. The organism is therefore shaped by them and modelled onto them. Now, it is natural for them to be meaningful things for an organism, so that their meaning is a phylogenetic product. Meaning, therefore, does not exist: only meaningful things exist. To be more precise, meaningful things do not exist either, but only "things that are meaningful for... (Prodi 1979: 188)

Natural, primordial, meaning is always a "meaning for" a certain form of life. Here, the influence of Jakob von Uexküll is evident. On the one hand, the living organism acts upon the world: "behaviours are not mere movements or tropisms, but they consist of perception (*Merken*) and operation (*Wirken*); they are not mechanically regulated, but meaningfully organized" (von Uexküll 1982: 26). Perception is always already an action, a doing, and a "reading" of the world. On the other hand, the terms of the relation—the organism and the "thing meaningful for" that organism—are linked together by what von Uexküll called the "functional circle" (von Uexküll 1982: 31; see also Brentari 2015: 107–15; Thure von Uexküll 1987; Kull: 1999). The most interesting aspect of Jakob von Uexküll's model is its circularity. The thing means something for the organism, but the organism cannot live without the thing. The "functional circle" (Fig. 4.2) completes the model presented in Fig. 4.1, by adapting it to the biological world. The primary effect of this model is that it effectively supersedes, by making it somewhat useless, the distinction between organism and signified thing. In fact, in a circle—and a "functional circle" is nothing but a circle—there is no beginning nor end. We are so accustomed to think in binary terms that we always conceive of a duality between a "subject" doing something to an "object". But von Uexküll's model overcomes this distinction, and Prodi takes this model very seriously. The distinctive characteristic of "the flux of living beings" (Prodi 1979: 14) is indeed its radical dynamic continuity: "it is necessary [...] *to think that mobility is the very 'substance' of nature*" (Prodi 1979: 15).

Fig. 4.2 The functional
circle of behaviour. (Jakob
von Uexküll 1982: 32)

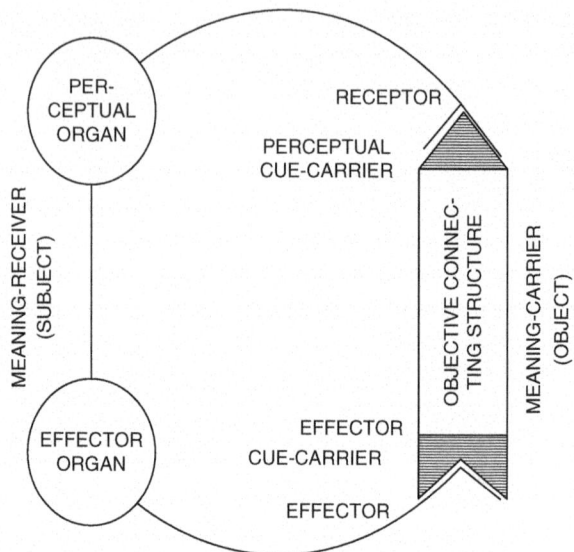

In this scheme the arrows of Fig. 4.1 transform themselves into a (functional) circle; it therefore becomes clear that Prodi's explanatory strategy is not simply non-
intentional and reductionist—a strategy that would attempt to dissolve or, more
precisely, to reduce complex phenomena to their simple constituents. It is not so
because, according to Prodi, while the ground of semiosis and of language is certainly in the world (that is, the material ground of semiosis is the world to whom the
signs make reference), such a ground is already intrinsically semiosic: that is, it can
be conceived as an infinite chain of translations and transformations of complex
assemblages into other complex assemblages. At all levels the world is made of
'meaningful' relations. Single organisms, for example, are "units of behaviour, of
reaction, of vital course (from life to death): but they are also modular realities, that
is to say, they are composed by many smaller units" (Prodi 1979: 17). As we have
seen for Prodi, nature is a "flux", continuity in change. This is the background
against which we must understand his concept of the organism's "reading" of the
world. When we read a book, for example, we jot down notes about what we are
reading, or we talk about it to a friend, or we can simply commit it to memory: in all
these cases, "to read" means to transform words into *other* words and thoughts (that
in turn will need further words-thoughts and so on). This is what Prodi means with
"reading". Just like a summary is a transformation of a text into another text, so
every "reading" is the transformation of a "meaningful thing" into another, a new
text, and so on. Life is precisely this "and so on".

To move from the line to the circle completely changes the approach to the philosophical and scientific problem of grounding. The ground is not that which should
be, in turn, grounded upon something else. Philosophy has long tried to identify an
absolute foundation: as we will see in the last chapter, with Prodi we find an unexpected solution to this problem, since human language—the most complex and

articulate form of natural semiosis—will be revealed as being grounded upon another language, intrinsic to nature. Linguistic semiosis is nothing but an extreme complication of natural semiosis. Prodi's explanatory strategy, then, cannot be represented with an arrow, whether this goes from the simple to the complex or vice versa. Rather, it is best exemplified by a circle, a figure without a basic, fundamental element. Prodi is well aware of this, just as he is conscious of the vain hope of attaining an absolutely original knowledge, a definitive foundation:

> in truth, no foundation is possible, because the ultimate operation of thought is a particular operation, like every other one. There is no chapter zero of knowledge, such that its employment would allow a glimpse upon the totality of reality, making the whole of ontology, for a fleeting moment, founded upon an enlightened gnoseology, instantaneous and exalted. Such an operation is completely illusory. (Prodi 1982: 6)

Prodi, then, abandons the idea of an ultimate foundation of knowledge. The relation between knowledge and the world— like that between semiosis and world—is characterized by continuity, as in the circle, where there is no beginning nor end, no before nor after. To hold the primacy of the circle means, once again, to prioritize the relation to the *relata*. Prodi calls this original relation "a tapestry of facts" that is a "tapestry or network of relationships" (1982: 8) accounting for how an organism can attain knowledge about (i.e. to "read") the world:

> A tapestry of facts must exist, of which knowledge is part and without which it could not operate nor could it be produced in the first place. [...] We can know things (and objects that can know things can be created in nature) because a) there is a given whole of relations; b) this whole is "traversable" with operations; c) it has produced systems that belong to the "tapestry", that is to say, systems included in the whole of relationships, and exploit factually existing conditions — being their manifestations. (Prodi 1982: 8)

The world is a "tapestry of facts", a whole of relations and not of isolated and individual things. As he writes in *Orizzonti della genetica*, the world is a vital "flux", without interruptions nor gaps. We will see below how Prodi, when tackling aesthetic and religious themes, seems to refer directly to the *Tractatus* (without explicitly quoting from it). In particular, his arguments seem to echo Wittgenstein's opening propositions "1.1 The world is the totality of facts, not of things; 2 What is the case, the fact, is the existence of atomic facts; 2.01 An atomic fact is a combination of objects (entities, things)" (Wittgenstein 1922: 25). But what does Prodi mean when holding that the world is a "tapestry or network of relationships"? He means to say that life is regulated transformation, that is to say "reading" or, which is the same, semiosis. This does not entail that the world is composed by signs (against pansemioticism see Hoffmeyer 2010: 603), but rather that the world *is* relation of relations. Indeed, semiosis means meaningful relation. And since a relation, as we have already seen, is only meaningful *for* an organism, to say that the world is a "tapestry or network of relationships" means that the living world is the whole of all biosemiosic phenomena: of all those "myriad forms of communication and signification observable both within and between living systems [...] representation, meaning, sense, and the biological significance of sign processes — from intercellular signalling processes to animal display behaviour to human semiotic artefacts such as language and abstract symbolic thought" (Favareau 2010: V).

According to Prodi this whole of relations—the world—defines the space of "material logic, which can be found in the facts, in the relations and among the elements of the horizon" (Prodi 1977: 43). This means, precisely, that the world itself is intrinsically logical-semiosic:

> we are used to linking logic only to the functioning of our thought capacities: but if those are present in nature, it is because they have differentiated themselves in nature, and since they act upon it their root is the same as those material exchanges that they are capable of interpreting. They are based on a logic of material exchanges, differentiated through a chain of increasingly more complex functions. [...] In this sense logic, at its deepest level, is a material tautology. What is, is logical. (Prodi 1977: 43)

Here the reference to *Tractatus* is explicit: a tautology is always true (like p ^ p). World and logic (semiosis) are one and the same. But what Prodi calls "material logic", i.e. the logico-semiosic clay of the world, immediately becomes "categorial logic" (Prodi 1977: 39). Because in the living world, every organism must be able to categorize its living world, that is, discriminate between meaningful and meaningless phenomena. Prodi's "categorial logic" is based upon selection, on the act of discrimination of that with which it is possible to entertain a relationship and that which does not afford such a relation. Such a logic engenders the first implicit categories ("edible" and "inedible", "sexual partner" and "non-sexual partner", "safe" and "dangerous", and so on). This "categorial logic [...] can be identified with biological organization in general" (Prodi 1982: 83), that is, with life itself. The world-life is then always and at the same time logico-semiosic.

Let us repeat that the key to the comprehension of Prodi's thought is his insistence on continuity. Prodi wants to offer an account of the possibility that from the world of things (where "material logic", articulating not yet meaningful relations between things, holds sway[2]), it is possible to arrive, through a chain of complications, to human language: the most complex semiosic system that exists. Prodi's crucial move for the resolution of this problem is his identification of the world of life with the world of semiosis. When Prodi talks of a "material logic", he is simply stating this point: the natural world is not a world of isolated things, but a world of relations, of regulated connections between things; for this reason "the event, and not the [isolated] thing, lies at the basis of material logic" (Prodi 1982: 27); indeed

---

[2] It is important to stress that Prodi did not place a neat separation between organic world (the world of semiosis) and the inorganic world, the world where there is not yet "meaning". Take the case of such a quotation from Thure von Uexküll: "The line drawn between organic and inorganic nature is not determined on the basis of random distinctive features, such as chemical makeup, size, complexity, or the form of the structure in question, but on the basis of a characteristic quality which can first be observed among living things and which is inherent even in the simplest forms of life, the protozoans. This inherent characteristic is the ability of an organism to react to stimuli, not just in a causal-mechanical way, but with its own specific reaction. From this point of view, all living organisms are considered autonomous, while the inorganic, including the tools and machines we use, remain heteronomous" (Thure von Uexküll 1987: 152). I think that Prodi would not agree with such a radical separation between what is "living" and what is "causal-mechanic". In fact, such a separation still seems to imply some form of dualism, and we know that Prodi wanted to get rid off of *any* form of dualism, even such a deeply rooted dualism between the living and the nonliving.

"the concept of a thing is inadequate to function as a starting point. Physics, study-ing matter in its constitutive parts, does not find entities that are intuitively definable as things" (Prodi 1982: 20). Ultimately "material logic" means nothing but relations and relations of relations. It is on these grounds that other forms of logic are "seam-lessly" (Prodi 1982: 15) developed: first the "categorial" logic and lastly "proposi-tional" logic, that of articulated human language.

Let us return to "material logic", the natural logic as a "tapestry of facts, a mate-rial tapestry" (Prodi 1982: 9). Since logic-semiosis coincides with the natural world, "the initial operation is therefore a preliminary identification of logic with onto-logic" (Prodi 1982: 16). Beyond this tapestry, this "network" is "traversable", that is to say it is knowable by a living organism, being the very same lifeworld that pro-duces "systems" capable of knowing it, systems that are nothing but "manifesta-tions" of this primordial network. The somewhat paradoxical outcome is that this world-life-logic assemblage achieves self-knowledge through its subsystems, through its own manifestations "since the beginning we therefore mean to interpret gnoseology as an internal function of ontology" (Prodi 1982: 9). Knowing becomes a function of being, a *partial* manifestation of being. Once again Prodi surprises us and unveils yet another possible hidden source of his peculiar way of philosophiz-ing—the theological model of Spinoza's *Deus sive Natura*—according to whom the same circularity between knowing (human animal) and nature (God) takes place:

> [h]ence it follows that the human mind is part of the infinite intellect of God; and therefore when we say that the human mind perceives this or that, we are saying nothing else but this: that God-not insofar as he is infinite but insofar as he is explicated through the nature of the human mind, that is, insofar as he constitutes the essence of the human mind-has this or that idea. And when we say that God has this or that idea not only insofar as he constitutes the essence of the human mind but also insofar as he has the idea of another thing simultane-ously with the human mind, then we are saying that the human mind perceives a thing partially or inadequately. (Spinoza, *Ethics*, II, XI, Corollary)

Every level of epistemic activity, from that of the virus to that of the scientist, is possible—it manages to reach onto the world—because it is always already within that very world (for this reason knowledge of the human mind, says Spinoza, can never be inadequate, since it is constitutively partial): "it is as if the network", the world itself, "would bend upon itself in certain areas, thus becoming capable of bet-ter reading and deciphering the things upon which it bends and comes to surround" (Prodi 1982: 41). Knowledge does not belong to the subject; it is the world that reads itself through the single organism, the single "reader". This is why it is impos-sible to find an absolute and definitive foundation to knowledge. Such knowledge would be possible only if a given organism could read the whole world from the *outside*. But "no observer […] can see the whole network", i.e. the entire world "nor can it say if the network is a whole. […] The observer is internal. There are no exter-nal observers" (Prodi 1982: 36). As we will see in the last chapters of this book, this intrinsic limit of natural epistemic systems implies important consequences of both ethical and religious nature. For now, it will suffice to insist that it is *logically* (and therefore biologically) impossible to escape the biological circle of knowledge:

[t]he terrain upon which we move — whilst looking for an explanation for this very movement — is a tapestry of existing facts. If a network of facts exists, and we are part of it, then
our nature (structure and functions) derives from it and is its own specification. It cannot be
neither contradictory nor external. We are never faced with the problem of having to justify
the network, but we need to be justified by it. […] This network is therefore continuous with
us, and not something extraneous which we could see as wholly external and confronting
observers. We hold it within us from the inside, because our epistemic capacities are simply
one of its organizational modes. Through the networked facts that compose us, and the facts
that connect us to the outside, we are seamlessly immersed in the wider tapestry of facts.
(Prodi 1982: 15)

Let us now try to reconstruct Prodi's thought as a whole. We began by ascertaining
that the world is a tapestry of relations, what Prodi identifies with the space regulated by the "material logic". But this is not really a beginning, since a relation has
always already begun—else it would not be a relation. The model of the circle
makes it impossible to identify an absolute beginning. We could say that at the
beginning, there was what already was. "Material logic" immediately becomes life
("categorial logic"). But life, as a construction and identification of natural meanings, is coextensive with semiosis: "it is clear that material logic and material semiotic coincide. If a material presence selectively interacts with another [categorial
logic]—unveiling it as a referent sign and operative trigger—then this would be a
logical condition, connected with the impersonal claim 'every time that…'", (Prodi
1977: 44). What is meaningful for a "reader" triggers a biological operation; such a
connection repeats itself every time the "reader" encounters the same "referent
sign", that is, the meaningful aspect of the world. Such a repetition transforms the
relation into a kind of "logical condition". In this sense it is possible to speak of
"material logic".

Life equals semiosis: "biology is natural semiosis" (Prodi 1987b: 147). But then,
applying the transitive property to this chain of equations, we reach the identity of
semiosis and world (since the world is coextensive with life) and then the final tautology (final but also initial, since in a circle every beginning is also an end and vice
versa): semiosis is equal to semiosis. this tautology means that semiosis is at the
origin of semiosis, and this simply means that the world (= semiosis = life) exists
and that everything that came before it was already world (= semiosis = life). It is
impossible to get out of the world.

Is this conclusion not a vicious circle? Since semiosis means relation, Prodi is
actually telling us that there was no beginning. More precisely, he is telling us that
at the beginning, there was already a relation—another way of saying that an absolute beginning never occurred. This is why Prodi holds that at the beginning, there
was the event, not the thing: at the beginning "we could say, 'there was change', if
only we weren't disposed to think of change as the change *of something* already
given" (Prodi 1982: 27). At the beginning there are "logical relations", and for this
reason, "the concept of a thing is not primitive" (Prodi 1982: 27). It can be seen how
Prodi always goes back to his core guiding ideas: the circle, continuity, and natural
semiosis. If relations are at the beginning, then things are nothing but "particular
kinds of logico-material relationships" (Prodi 1982: 28). Hence the evolutionary
equation—at the beginning of semiosis (i.e. of meaning), there is natural semiosis

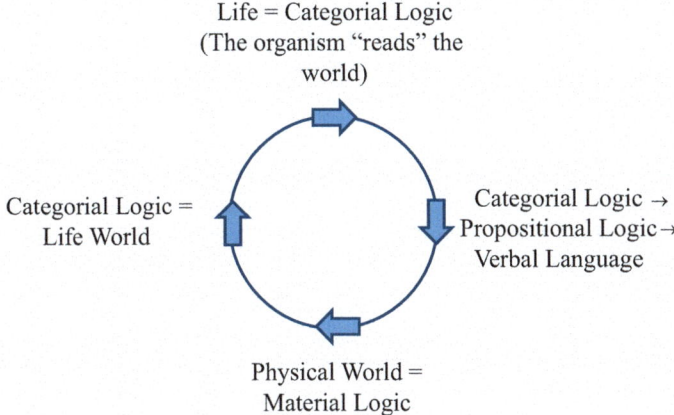

Life = Categorial Logic
(The organism "reads" the world)

Categorial Logic =
Life World

Categorial Logic →
Propositional Logic →
Verbal Language

Physical World =
Material Logic

Fig. 4.3 The natural world as a whole of relations: material logic, categorial logic, and propositional logic

("material logic")—is not an empty tautology. Prodi is arguing that the world is this relational continuity, a relational circle wherein every moment presupposes and develops the previous moment. It is crucial that this figure should be a circle, since only this model allows us to understand how a virus can develop into a scientist and a philosopher. Essentially, Prodi is uniquely concerned with this question: how is it possible to arrive at a scientist or a philosopher starting with a virus? In Fig. 4.3 I tried to schematically summarize this chain of identities (the arrows represent the direction of the process, considered as both an evolutionary and a foundational model, see Cimatti 2000b).

What happens when from propositional logic—the core of human language—we arrive at material logic? That is, when language refers to the world? This is more a return than a point of arrival: language can *say* the world because, in the final analysis, language is simply the world saying the world. It is always the same world, albeit manifesting itself in different ways: "man […] is nature thinking itself, [it is] the interiority of nature" (Prodi 1987b: 93). This statement already alludes to the internal dynamics of this circle, which is not closed onto itself (if it was, language would be useless, since there would be no necessity to express what one already knows) because such a self-consciousness of nature is continuously expanding. That is to say that this movement—no longer a circle but a spiral—(see Chap. 10) will give access to ever larger parts of previously unknown regions.

If we return to Fig. 4.3, we realize how paradoxical and unfashionable this model appears today. According to it, the great dream of contemporary analytic philosophy is unattainable: to naturalize the mental, the project feverishly pursued by many contemporary philosophers and scientists (see, e.g. Millikan 1984; Dretske 1995; Papineau 2003). For Prodi, language—and semiosis in general—is not something distinct from the world of life nor is it set against nature, biology, or matter. Prodi does not think it necessary to simplify complexity but rather to show how what we thought as simple was always already complex. This ultimately means to completely get rid of the distinction between simple and complex. If life is *already* semiosis, it is meaningless to puzzle about the correct placement of the border between nature and culture. We should rather try to understand *how* nature becomes culture and—and this is a *far more* interesting question—how nature is *already*, somehow, cultural:

> [t]he duality cannot be overcome by synthesis, but rather through the acknowledgment that, at bottom, there is no duality at all. Penetrating deeply into the study of nature we can see how, on the one hand, in its most sophisticated and recent regions it becomes moral. On the other, going back towards its origin, we can see how every form of knowledge is a commonality with things, it is participation. So it is for man. Knowable reality is very different from what is usually presented to us: things are mute, but they can respond when suitably interrogated. By doing so, a fundamental transaction with the instruments used to probe it — that is to say, with us — can be revealed. Our knowledge derives from more ancient forms, and goes all the way back to the root of the biological, which appears to us as "elementary knowledge", ever since the very first steps of its organization. (Prodi 1987b: 119)

Finally, it is for this reason that for Prodi every kind of dualism, epistemic and semiotic (that is to say the question of the relation between form and content, sign and reference, or, more generally, subject and object), literally vanishes. His circular model is developed as an image of continuity, the same continuity that links all living beings together. For this reason—since a clear split between knower and known, internal and external, is never given—the world's primary characteristic is its intrinsic knowability:

> [an] organism constitutes itself because reading some external meanings grants it some advantages, it really *constitutes itself onto them*. It is necessarily complementary to these external things, since they are what the organism can read, and with whom it can selectively interact. There obtains, therefore, a relation of complementarity and of adaptation between an organism and reality, because the reader builds itself onto its reality (a given reader constitutes itself onto a given reality). [...] An organism knows/interprets (has a specific relationship with) the reality *onto which* it has constituted itself. It interprets the world through its own categories, but these categories have been constituted by *that* world itself. *An organism interprets its own genetic area.* [...] The organism knows the reality that constituted it. Things let themselves be known, because they have constituted the interpretive categories needed for their knowledge. (Prodi 1987b: 143–144)

Chapter 5
The Biological Model: For an Anti-Cartesian Semiotics

We think that the threshold for "sign" is situated at the very beginning of the biological domain, characterizing its origin and its basic structure [...] Life begins when to such a uniform world, conditions of selectivity are superimposed or, better, when conditions of selectivity are generated from the conditions of uniformity. [...] An enzyme, which can be considered the simplest example of this status, selects its substrate among a number of meaningless molecules with which it can collide: it reacts and forms a complex with only its molecules partner. This substrate is a sign for the enzyme (for its enzyme). The enzyme explores reality and finds what corresponds to its own shape: it is a lock which searches and finds its proper key.

(Prodi 2010: 329)

Abstract For Prodi, the fundamental semiotic interaction obtains between two molecules. The first selectively "reads" some superficial characteristics of the second, allowing it to establish a link. The fact that a link between two molecules is possible makes the second molecule "meaningful" for the first (and vice versa). "Natural meaning" thus arises. At the beginning of semiosis, there is a selective material operation, wherein a certain material configuration is "preferred" to another. All the other forms of semiosis derive from this fundamental operation. Prodi's theoretical proposal, then, does not presuppose the existence of any intentional process. Semiosis, Prodi argues, does not need a subject or any psychological intentionality. Consequently, semiosic processes are completely natural and are not an exclusive prerogative of human beings.

Keywords Natural Semiosis · Descartes · Peirce · Subject · Communication

A semiotics can be said to be Cartesian—either explicitly or implicitly—if it is grounded on the notion of the "subject" and on concepts linked to it, for example, "communicative intention", "purpose", and "convention". Therefore, a Cartesian

© Springer Nature Switzerland AG 2018
F. Cimatti, *A Biosemiotic Ontology*, Biosemiotics 18,
https://doi.org/10.1007/978-3-319-97903-8_5

semiotics presupposes, often without realizing it, an original dualism between the mind and body, between signifier and signified, and between internal and external. A first consequence of this kind of semiotic theory is that semiosis and language are considered to be exclusively (or primarily) human phenomena, thus excluding the possible existence of forms of semiosis (or in general of mental activity) in non-human animals. I will formulate three examples of a Cartesian semiotics: the first is implicit in Claude Shannon's communicative model, the second is more properly semiotic, and the third is taken from the field of psycholinguistics. I should stress that I am using the adjective "Cartesian" not in a historical sense but rather as indexing a certain attitude privileging subjectivity (from this point of view, most contemporary Cartesians are not aware of being Cartesians and in fact often think of themselves as doggedly anti-Cartesian). These are not particularly recent examples, since what matters here is to delineate the general outline of a certain attitude towards semiosis, according to which semiosis pertains to the intention of communicating. I use these examples because they present such attitude in a particularly clear manner and allow us to comprehend—as a contrast class—Giorgio Prodi's point of view.

The first example is the most surprising one, since it is hard to imagine Claude Shannon, engineer and mathematician, as a Cartesian. And yet his celebrated model of communication is intrinsically Cartesian, that is to say dualistic. In fact, any semiotic model that construes semiotics as originating from a mind—what Shannon calls "information source", a seemingly neutral term—can be defined as Cartesian.

As it often happens, a picture paints a thousand words. According to this diagram, communication can be modelled as an arrow proceeding from a source towards a destination. Everything begins, somewhat magically, in the first box—the information source. The diagram does not tell us what happens inside that first box, but it does clearly show that everything begins there. According to this model—so pervasive today that nobody remembers that it is, in fact, just a model and not the fact it purports to explain—communication has an absolute beginning in the information source only to then flow, relatively undisturbed, towards its destination. The radical (yet perhaps unintended) dualism in the Shannon-Weaver model is explicit. In the first line of their book, indeed, they write that "the word 'communication' will be used here in a very broad sense to include all the procedures by which one mind affect another" (Shannon and Weaver 1949: 3). Communication begins with a *mind*. Although Shannon and Weaver continue by specifying that this model can also be applied to automatic systems, it is clear that the "information source" has the mind's most important characteristic: it selects what is worth communicating. As they write, "the information source selects a desired 'message' out of a set of possible messages" (Shannon and Weaver 1949: 7). What is a Cartesian mind if not a free capacity for desiring and choosing? The problem with this model is that, clearly, it cannot be naturalized: it is meaningless to talk about "desire" or "choice" when referring to viruses or enzymes. If semiosis really worked according to the model explicated by Fig. 5.1, then it *would not* be a natural phenomenon.

Let us move to another example, that of the semiotician Luis Prieto, very active during Prodi's time and, as a member of his same semiotic circles, certainly known

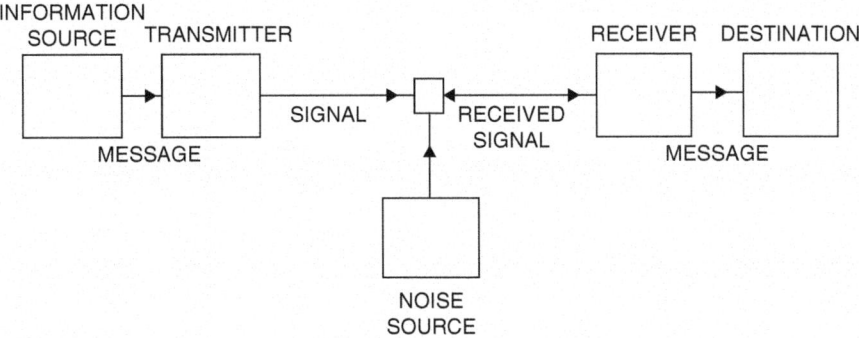

Fig. 5.1 The "standard" model of communication. (Shannon and Weaver 1949: 7)

to him. Prieto had no doubt that the ultimate and irreducible foundation of semiosis is the subject, since only the subject can operate the choices that institute a practice, and precisely a semiotic practice.

> [t]he body of the subject, thanks to the capacity of choice it is endowed with, has a peculiar characteristic: it can be a cause — for example the origin of a practice — *without in turn having been the effect of anything.* The existence of a practice requires that [...] the transformation of the body of the subject — in turn causing the transformation of the object that in the practice comes to play the role of means — be a *decision*, that is to say *a non-natural transformation.* (Prieto 1991: 17)

For Prieto—a Cartesian to the bone—semiosis begins with a "non-natural transformation". It follows that the fundamental characteristic of the semiotic subject is its being a *cause* without *having been the effect of anything.* If one has to be a Cartesian, it is better to be Cartesian all the way down. Prieto, then, conceives of the subject as a kind of God, since the classic definition of God is, precisely, to be a non-caused cause. Thus, the subject of semiotics (Shannon and Weaver's "information source") like that of Descartes becomes—whether or not this was Prieto's intention—something profoundly different from every other thing in the world, all those things being the effect of something else. Isn't this a perfect example of a dualism? This is the premise that, for Prieto, underlies semiosis: the only presupposition capable of, for example, justifying the distinction between signal and symptom—the first being voluntary, unlike the latter. (In those years Eco was promoting, among others, a system of classifications of signs according to the "degree of consciousness of their emitter" [Eco 1973: 38].) The limit of this kind of explanation for semiotic phenomena is that, in fact, it fails to explain anything at all. If, in order to explain the existence and use of a sign, it is necessary to postulate the existence of an entity—the semiotic subject—which would be by definition non-natural, this clearly would not be a naturalistic or biological explanation of semiosis. If, in order to explain a fact, it is necessary to expect some kind of "miracle"—and the semiotic subject who, unlike other material things, withdraws itself from all causal chains, indeed would be a kind of miracle—then such an explanation does *not* explain anything at all.

The third example comes from the work of the psycholinguist Willem Levelt, particularly his important—and appropriately titled—monograph *Speaking: From Intention to Articulation* (1989). Levelt examined only one kind of semiosis: the phenomenon of articulated spoken language. However, the way in which he conceives of the subject is thoroughly Cartesian, that is to say, dualistic. Commenting on a model that is nothing but a more complex version than the one presented in Fig. 5.1, Levelt writes: "talking as an intentional activity involves conceiving of an intention" (1989: 9). The starting point of Levelt's model is yet another box, which he calls "conceptualizer". All the processes that matter take place inside this closed box, without an explanation: the "conceptualizer" produces a "preverbal message" (1989: 9) that serves as a foundation for the subsequent steps in the process of linguistic coding of the message. The obvious question is where does this "preverbal message" come from? Levelt offers no explanation, and indeed he cannot explain it because his, like all Cartesian models, is grounded on an absolute foundation—the consciousness of the subject. Such consciousness cannot be explained: indeed, it functions as an unshakable ground, supporting the entire explanatory edifice. Levelt realizes that there is a *huge* problem here, but he wriggles out of it the easy way: by leaving some issues to be resolved by future research. Indeed, by postulating the existence of a system like the "conceptualizer", Levelt is "in full awareness that this is a reification in need of further explanation. We are, of course, dealing with a highly open-ended system involving quite heterogeneous aspects of the speaker as an acting person" (1989: 9). Summing up these three examples, we can say that a semiotics is Cartesian if it presupposes, at the starting point of the process of semiosis, the existence of a mind, an uncaused cause, or a preverbal message. In all these cases, the process of semiosis begins with a non-natural entity, a postulated and unexplainable entity. It will now be clearer why, in the previous chapter, I have insisted on the difference between the model of the circle as compared to that of the line (or arrow): the latter is a dualist and discontinuous model, the former a biological and continuous one.

In opposition to the examples I have just offered, from the beginning of his work, Prodi champions a radically anti-Cartesian approach[1] to semiosis: the method of semiotics is "the construction of linguistic models starting 'from biological models'" (Prodi 1983b 173). The Cartesian and intentional semiotic model postulates a

---

[1] The standard model for an anti-Cartesian semiotic is offered by Peirce in his "Questions Concerning Certain Faculties Claimed of or Man" (1868). For similar reasons, Prodi is also a critic of phenomenology, since it presupposes the ability to reach an original and autonomous state of consciousness. Prodi, in particular, doubts the very existence of the act of ἐποχή or "bracketing". According to Edmund Husserl, the phenomenological stance implies the ability to suspend judgment regarding the general or naive philosophical belief in the existence of the external world and thus to examine phenomena as they are originally given to consciousness. On the contrary, according to Prodi "there can be no 'bracketing' of things, especially so when it comes to that thing (the knowledge-structure) which has organized itself onto things by perceiving and manipulating them; a thing which is nothing but a complex 'linking-thing' [*cosa di collegamento*]. Things are always integral and necessary to each other, that is, they are reciprocally constitutive at every step of the process which tries to separate them; the darkness is populated by their presence" (Prodi 1974: 16).

mind as an explanatory principle: such a mind would fix the conventions that will later crystallize into codes and decides to use something as a sign for something else. Prodi's biological and evolutionary model, on the contrary, aims to explain how mind can emerge out of the non-mental and how semiosis can emerge from non-(explicitly) semiosic interactions. The semiotic model, then, is primarily synchronic and hierarchical, precisely because it places the mind at the apex of its explanatory system. Conversely, the biological model is diachronic—appealing to evolution in order to comprehend synchrony—and horizontalist or circular, since it attempts to explain complexity as the global result of the non-intentional interaction of a multitude of simple agents, each of which is merely pursuing a local and defined objective (see Minsky 1985; Dennett 1995; Parisi 1999).

From Prodi's perspective the semiotic subject—by definition separate from the world of natural things and events—becomes a "reader or interpreter", that is to say a material entity with a natural capacity to enter into a relation (merging) with certain objects. This point needs repeating: when Prodi talks of a "reader or interpreter", he is not thinking that the organism would be capable of reading the world without any constraint. The organism *qua* reader "reads" the world by seeking complementary things, objects with which it can establish stable and meaningful (i.e. biological) links. For Prodi the lifeworld, we need to remember, coincides with the world of categorial logic (in turn, a transformation of the original material logic), that is to say the world of interactions and mediations: "we can [...] identify the space of signs with that of biology, and interpret this as a reservoir of systems of symbols" (Prodi 1977: 48). Within this network, every "reader" (really a stratified assemblage of simpler readers) can only enter into a relation with a small subset of other assemblages; that is to say, every organism (every unitary assemblage of readers) "categorizes" the surrounding world by individuating (classes of) things with which it is possible to interact, setting them apart from those with which interaction is impossible (a vastly larger class). This "reading" operation is not to be interpreted as an intentional or arbitrary process; certain configurations of proteins, for example, can only interact with determinate assemblages—their possible "merging" with other assemblages is conditioned by their material characteristics *and only by those*. Such an assemblage "reads" the surrounding world, all the things it enters in contact with: following Jakob von Uexküll, *the environment* of each living organism is nothing but the sum total of the things it can interact with, plus the complementary set of things with which interaction is impossible. Prodi's "reading", therefore, is something that takes place *before* the emergence of consciousness and intentionality. Besides, Prodi uses this particular term in order to stress how semiosis is not an element added to the natural world from the outside. The world of life is always already coextensive with the world of semiosis, and "reading" too can therefore be nothing but a natural phenomenon. Semiosis, at first, is a selective physical contact between things. At the very beginning, there are:

> [t]wo interacting material objects (coming into contact with each other [...]) They must be able to be transformed by this contact, subordinated to a precise condition of correspondence. That is to say that change takes place only when two objects somehow have a steric correspondence [i.e. they coincide tridimensionally]. The concept of specificity, then, finds

its material equivalence into that of spatial correspondence, of reciprocal adaptation, and of complementarity,[2] roughly exemplified by a condition of "correct fit" or interlocking. (Prodi 1977: 22–23)

As it is always the case, the *relation* comes first—in this case represented by steric complementarity (linked to the special distribution of the atoms in a molecule). Only subsequently, and thanks to their relation, the possibility of individuating *two* things that entered into a relation becomes actual: the reader—the thing that categorizes the world and selects "its" vital environment—and the read thing, that is to say the entity the reader can interact with. It is important to note how these two entities are fully complementary and located on the same level, such that the relation between the two can also be seen from the point of view of the "read" one, as selecting "its" reader. Every reading operation constitutes a new assemblage which, in turn, will trigger new reading processes of its environment and thus will progressively assume the role of an ever more complex object, with increasingly articulate reading skills.

> [t]he existence of the [reader] profoundly modifies — yet without introducing any new principle — the relationships in the network, since in this network appears now a central node (and obviously many decentred nodes). A logico-categorial reader, acting as a selector, is a convergence point. This does not imply any kind of hierarchical "centrality". [...] This reader evolved out of things themselves, and moves efficiently upon them precisely because it originates from them, and knows them: this makes its logical position even more peculiar and new. The things do not converge upon the reader (it has no privilege, does not

---

[2] In *Kant and the Platypus*, Umberto Eco defines the concept of "primary iconism" in explicit reference to Prodi, mentioning the semiotic domain of complementarity—"the icon is the natural willingness of something to *correspond to* something else"—and finds in it the ground for "superior cultural phenomena". Eco explicitly mentions Prodi's *Le basi materiali della significazione*: "in no way am I repudiating the distinction (which remains fundamental) between signal and sign, between dyadic processes of stimulus-response and triadic processes of interpretation, so that only in the full expansion of this last do phenomena such as signification, intentionality, and interpretation (however you wish to consider them) emerge. I am admitting with Prodi (1977) that to understand the higher cultural phenomena, which clearly do not spring from nothing, it is necessary to assume that certain 'material bases of signification' exist, and that these bases lie precisely in this disposition to meet and interact that we can see as the first manifestation (not yet cognitive and certainly not mental) of primary iconism" (Eco 1999: 107). However, Eco does not seem to have moved on from his previous stance (as, e.g. in Eco 1976) since he still sets this dyadic iconism apart from "triadic processes of interpretation". Conversely, the radicality of Prodi's proposal lies precisely in its questioning of such a separation, considering all semiotic phenomena intrinsically dyadic (reducible to chains of dyadic links). Consequently, this means that notions like intentionality (which is a triadic entity) can be abandoned. It is no coincidence that Eco defines "primary iconism" something that, for Prodi, is not at all iconic (i.e. it is not properly a sign). The point of this discussion is not so much how to assign to the mental-triadic sign a hook in the world (this, for Eco, is the function of "primary iconism"), rather the point is to relinquish the unreflective presupposition of Cartesian semiotics, grounded on triadic relations (Peirce's Thirdness). However, while in *A Theory of Semiotics* Eco places dyadic phenomena below of "the lower threshold of semiotics", in *Kant and the Platypus*, he is more sympathetic to Prodi's ideas: "[y]et again I would refrain from using terms such as 'sign', but it is beyond doubt that when we come up against this lock that seeks its own key, we come up against a protosemiotics, and it is to this protosemiotic disposition that I would tend to give the name of natural primary iconism" (108–109).

attract things): it is a purely operative convergence. They face a situation that traps and selects them. The reader, then, has the character of a unitary term of a comparison, and since it flows out of the things themselves, it is as if they delegate it. The terms might be misleading: we should remember that it is not an intellectual reader, as it would seem natural to believe, but rather an operative reader. (Prodi 1982: 52)

Prodi's rather ambitious objective is that of radically biologizing semiotics. In order to comprehend what this choice would entail, we can start considering Peirce's classic triadic definition of a sign: "[a] sign stands *for* something *to* the idea which it produces, or modifies. Or, it is a vehicle conveying into the mind something from without. That for which it stands is called its ***object***; that which it conveys, its ***meaning***; and the idea to which it gives rise, its ***interpretant***" (CP: 1.339). The sign is composed of three elements, sign, object, and interpretant, and in this sense, it is perfect Thirdness. But "Thirdness [...] is the same as mediation" (CP: 1.328). So, the sign is mediation. To biologize semiotics would rather mean that the primordial sign, the biological one, is without mediation. Prodi then stresses the need to dissolve the triadicity of the sign into extremely complex chains of dyadic relations, complementary between reader and environment.

As we have seen the most direct inspiration for this line of reasoning are Charles Sanders Peirce's three ceno-Pythagorean categories: Firstness, Secondness, and Thirdness (Perice, CP: 2.87 and 5:66). Here we encounter a nontrivial problem: if, on the one hand, Prodi borrows Peirce's anti-Cartesian orientation of thought, on the other his radical attempt to biologize semiotics ends up deflating the role of Thirdness and Firstness, in favour of Secondness, the category that, more than the others, exemplifies the intrinsic relationality of *natural* semiosis. For Prodi, "biology as a linguistic-connective function, and language as an advanced development of the whole of biological organization (as well as the underlying logic) [are] the most sophisticated expressions of biological organization" (Prodi 1983b: 173). Human semiosis—i.e. "superior cultural phenomena" (Prodi 1983b: 183–184)—is not an a-historic and unexplained result of the subject's free and voluntary activity, but rather the final outcome of a real life, always already semiotic. In this frame, "the very elaborate linguistic function operative in man, used for complex communication, represents the evolution of other communicative functions, in turn derived from the evolution of non-connective/linguistic functions. Biology, seen from this perspective [...] is the study of the development of such a linguistic function" (Prodi 1983b: 179).

Prodi's radical biological approach is an a priori rejection of any kind of dualism. In this sense Prodi, although acknowledging his debt to Peirce, goes beyond Peirce's semiotics. According to Prodi Peirce is unable to explain the material genesis of semiosis. Although Peirce's semiotics is explicitly anti-Cartesian, an unexamined remnant of dualism remains present in his thought:

Peirce does not need to postulate the intentionality and conventionality (i.e. the artificial character) [of semiosis]: however, in the way he articulates the problem of semiosis, the sign is something already given as a mediator, already part of a semiotic function, the genesis of which remains completely obscure. It is therefore necessary to go beyond: not

simply to abolish intentionality, but — at the most basic stage of the process of signification — to abolish mediation itself. (Prodi 1977: 158).[3]

The sign, for Peirce, is mediation and representation. For Prodi the sign—at least in
its initial phases—is *not* mediation: "the sign is not something that holds the representation [...] of something else; it is a natural thing that corresponds to (is a function of) something else" (Prodi 1977: 158). To truly challenge Descartes means not
only to eliminate intentionality, but also (and most importantly) to do away with
mediation. The natural sign is a "thing" that "doesn't send back towards an indefinite chain (a sign explained by a sign, explained by another sign and so on) but a
chain with a finite number of interactions" (Prodi 1977: 158). The model of natural
semiosis is that of the complementary relation between DNA and the amino acids
that compose proteins. In this example no mediation takes place: on the contrary, to
every sequence of nucleotides corresponds a specific amino acid. This is the natural
genesis of semiosis: "the correspondence rules between DNA and proteins [...]
represent the most conspicuous and general example of this historical interpretation
of meaning, what I have called natural semiotics, upon which the whole of biology
is founded" (Prodi 1989: 36–7).

Perhaps a Peirce theory exists that is closer to these radical ideas of Prodi, for
example, Peirce's "Evolutionary Love" essay (1893), where the American philosopher writes that "all matter is really mind" (CP 6.301) or again "matter is *effete*
mind, inveterate habits becoming physical laws" (CP 6.25; from the Latin *effēta*,
"exhausted"). Prodi could interpret this claim not as a form of idealism (very far
from his thinking, but not from Peirce's, who indeed explicitly talks of an "objective
idealism") but rather as an alternative formulation of the thesis that life is intrinsically relational, that it is "categorial logic" (as Prodi calls it), and that for this very
reason—because it can never be pure and free from links with other things—it can
evolve into a semiosis. Prodi's integral semiosic materialism is, in fact, not so far
from Peirce's cosmological theories. For example, when Prodi writes that "life is
the never-ending imperative for the research of meaning, and it long predates human
reason" (Prodi 1989: 94), he is really claiming that life is also mind, because without a mind there would be no meaning and in particular no "search for meaning". In
this case "mind" does not mean, as it would according to a Cartesian model, intentionality and consciousness, nor does it imply any form of dualism. On the contrary,
it means absolute dyadic relationality. In fact, the cosmological speculations of the
late Peirce seem mostly to focus on the theme of continuity, a central theme (as we
have seen in Chap. 4) in Prodi's thought as well. Indeed, for Peirce to hold that "all
matter is really mind" means that matter is Thirdness too: "[i]f you take any ordinary triadic relation, you will always find a *mental* element in it. Brute action is
secondness, any mentality involves Thirdness" (CP 8.331). Leaving aside Peirce's
definition of Secondness (which Prodi interprets as *the* fundamental semiotic category), the American philosopher is here saying that, on the one hand, "[t]he thread

---

[3] A possible source for this radical stance of Prodi's might have been Mead. See, for example,
Mead 1922.

of life is a third" but, on the other, also that "[c]ontinuity represents Thirdness almost to perfection" (CP 1.337). Semiosis means mind, mind means Thirdness, Thidness means life, and life means—and this is Prodi's objective—*continuity* (on the theme of continuity in Peirce, see Parker 1998; Fadda 2013).

In a way, this is a paradoxical conclusion, since the category of continuity seems to be much closer to Secondness than it is to Thirdness, and this confirms that Peirce's "Thirdness"—a category that unfolds fully in semiosis—is really a kind of Secondness, as if looked from a distance, so to speak, that is to say without paying attention to that infinity of steps linking every entity with every other one: "if we could take a sign apart to its constitutive steps we would discover [...] that every step is in turn composed by a reader-thing reading another thing, which in turn becomes for it a sign, and gets read by another thing, and so on" (Prodi 1987b: 147). From this point of view, it seems correct to claim that Prodi does not reduce semiosis to "Secondness" as much as he shows how the Secondness of the living world is not comparable to a simple causal interaction (it is not, as Peirce put it "brute action")—it is rather selection and discrimination. If Prodi is cautious of distancing himself from de Saussure, who restricted semiosis solely to the human sphere, "his" Peirce is however a "lowered down" Peirce, so to speak, whose conceptual apparatus is completely reoriented in order to account for phenomena that the American philosopher never took into consideration (even though, I believe, he would not have considered extrinsic to, or intractable for, his understanding of semiotics). So Prodi writes:

> [t]he demarcation of the field of semiotics is a crucial point. According to de Saussure's foundation semiotics is the science of artificial and conventional signs, like language and other rule-bound systems of inter-human communication (like for example rules of politeness, traffic laws, military signs, and so on). From this point of view Peirce characterizes a generic situation, not necessarily a human one, since the process of semiosis takes place whenever a mediation between an interpreter and a thing — by means of an interpreter — obtains. But in Peirce's framework [...] the only possible domain for this semiosic process is a human one, or at least the act of interpretation is always configured as anthropomorphic and anthropocentric. (Prodi 1977: 158; see also Fadda 2014)

The goal, then, is to find in Peirce the resources to move beyond Peirce, particularly leaving behind his unreflective "anthropomorphic and anthropocentric" prejudice. A useful concept for pursuing this project is that of "habit", a concept employed by Peirce for more than just humans or living organisms (West and Anderson 2016). Indeed, for Peirce, "[a] habit is not an affection of consciousness" (CP 2.148). Consequently, material items too manifest "habits", that is to say they follow a certain pattern of behaviour: "the existence of things consists in their regular behavior" (CP 1.411). What else is a "habit" if not a kind of "regular behaviour"? And for this reason "[w]hat we call a Thing is a cluster or habit of reactions" (CP 4.157). A "habit" is therefore the sum total of the relations into which a thing can enter. A thing is *nothing but* this "cluster" of relations. For the most part, this applies to organic things, particularly protoplasm: "protoplasm has its active and its passive condition, its active state is transferred from one part of it to another, and it also exhibits the phenomena of habit" (CP 1.939). In general, a "habit is by no means exclusively a mental fact. Empirically, we find that some plants take habits. The stream of water that wears a bed for itself is forming a habit" (CP 5.492). His rein-terpretation of the concept of habit as a relation, and therefore as a form of natural semiosis, allows Prodi to overcome the anthropocentric prejudice he still detects in Peirce's work.

Let us take the extreme example of a freshwater body. In its flowing, this body "knows" a certain underlying substrate of rock and soil, such that this repeated action—a thoroughly non-intentional and unconscious action—"digs" the riverbed, that is to say its "habit". How would Prodi interpret this phenomenon? There are two "things": the flow of water and the soil upon which it flows (to be precise, the two things do not exist in isolation; we distinguish them only in order to make Prodi's reasoning more explicit). The first thing—the flow of water—"chooses" certain characteristics of "its" environment, for example, gravity and the elevation profile of the landscape it traverses. Similarly, the riverbed "chooses" the liquid nature of the first thing ignoring, for example, its colour. It is therefore established a relation of complementarity and reciprocal adaptation between river (in its initial development) and terrain (in its becoming a riverbed): "the correspondence between organism" — (although, as in the case of a river, not necessarily an organism) — "and its epistemic correlates or things (that is to say the situation we labelled 'adap-tation' or 'being made for') is the outcome of evolution, which produced different kind of relation for different kinds of complexity" (Prodi 1979: 182–183). Prodi here speaks of an "organism", but there is no reason to restrict his reasoning to bio-logical organisms alone. Just as the river adapts itself to the shape of the terrain it crosses, so the terrain models itself under the flow of water. We are witnessing an "always *in fieri* situation of correspondence with its own intrinsic and immanent logic" (Prodi 1979: 183). For Prodi this radical immanence is nat*ural semiosis*. The riverbed is shaped by the flow of the river, and the river is shaped by the character-istics of the topography it encounters: "complementarity gets [reciprocally] consti-tuted on the object read" (Prodi 1979: 183). The converse is also true: it should not be forgotten that Prodi's explanatory model is the circle, not the line.

For Prodi, Wittgenstein *Tractatus* is another important theoretical point of refer-ence. This might seem surprising. Prodi refers to the vertiginous chain of equiva-lences used by Wittgenstein to link thought with the world: starting with proposition 3 of the *Tractatus*: "[t]he logical picture of the facts is the thought" and proposition 4, which explains how "[t]he thought is the significant proposition". Following this, Wittgenstein defines language stating that "[t]he totality of propositions is the lan-guage" (4.001). The first step of this chain of reasoning, then, is the equivalence between language and thought. Moving to the second step, we are told that "[t]he proposition is a picture of reality" (4.01). But how does a picture function? Wittgenstein explained that in proposition 2.161, "[i]n the picture and the pictured, there must be something identical in order that the one can be a picture of the other at all". Assembling these proposition we reach the conclusion that language (= meaningful propositions = logical picture of the world) is a kind of manifestation of the world itself. Indeed, in proposition 5.7711, Wittgenstein writes that "[t]o give the essence of proposition means to give the essence of all description, therefore the

essence of the world". This last step in the reasoning has a curiously Prodian fla-
vour: the world is not an inert collection of isolated items, but a living world, since
"[t]he world and life are one" (5.621).

Let us return to Prodi. The natural world has always been (as Wittgenstein held
too) a "tapestry of facts", of relations and connections. In his *La storia naturale
della logica* (1982)—an indispensable text to comprehend Prodi's entire project—
this "tapestry of facts" coincides with what he calls "material logic". The world
coincides with this natural logic, because the world is made of events, i.e. of relations. Prodi rejects the traditional conception of nature as simple matter and as a
pure object facing a subject which, from the outside, would shape it and assign to it
an arbitrary form. Such a nature-object would be all the more passive and inert, the
more the subject functions as sole actor of order and action. The Cartesian semiotic
model, focused on the subject, both presupposes and confirms this radical dualism,
with the additional consequence of making it impossible to resolve the relation
between subject and object, on the one hand, and that between sign and meaning, on
the other. The very possibility of human language, so radically different from any
other form of semiosis, becomes then mysterious: How can something so complex
and stratified have emerged out of nothing? It is no coincidence that in the Cartesian
model, the "solution" of this problem consists (as we have seen in Prieto's case) in
proclaiming the subject capable of taking a primordial decision, a "non-natural
transformation".

Prodi's biological model, on the other hand, tries to tackle the semiosic problem
from a different angle: instead of seeing semiosis as something "invented by" a
human subject, or a community of subjects, turns the semiosic problem into a biological problem. Indeed, to employ the circle as a model for semiosis means to
choose continuity between different forms of life as a guiding principle (as we will
see below, this does not entail a neglect of species-specific differences between different forms of semiosis):

> the *continuum* between things and interpreter, between nature and culture, the noumenon
> and its semiotic-phenomenal correlate is the foundation of knowledge, and is expressed by
> saying that the reader is derived from his reading world. [...]. In substance, to communicate
> does not mean to intervene into extra-semiotic circumstances, but rather to immerse our-
> selves into a world that is always-already semiotic, and that has generated us as readers.
> (Prodi 1977: 164–165)

So, the development of systems able to categorize their environment—categorial
logic—starting from the intrinsic connectivity of the natural world (the space of
material logic) evolves in the human animal, into a propositional logic through
which:

> organisms can have an epistemic contact with both reality and the ways of seeing this real-
> ity. Knowledge knows, at one and the same time, both the things and its own processes: it
> self-sustains itself, and it is able to see beyond its immediate and categorial terms, towards
> its farther, genetic conditions. It can also reflect on its material and categorial genesis. On
> the grounds of this familiarity with external terms, it can modify reality according to a
> project. (Prodi 1982: 84)

Clearly, this is *not yet* a theory about the origin of human language. At best, it suggests where one should look for a solution to this question and in particular where not to look: neither in the consciousness of a semiotic subject (a Cartesian, top-down, kind of solution) nor in the forms of communication of other animals (the classic evolutionary bottom-up solution). The first is simply not a solution at all, since it merely proclaims the specialness of the human against all other living beings; the second solution, on the other hand, does not capture all those distinctive characteristics of human semiosis: the latter is clearly linked to other forms of natural semiosis but still very different from them. Both solutions are linear explanations, and we know how Prodi privileged solutions based on the model of the circle. For Prodi the specificity of human semiosis, of verbal language, is its being an adaptation to communication itself. While the bee dance is adapted to the environment of the beehive, the species-specific semiosis of human animals is, for Prodi, adapted to language itself:

> [w]hich selective model can be conceived for the phylogenetic determination of human logical-linguistic competence? The topic is a difficult one: we do not have any documents to go by, and necessarily we have to speculate. […] We can imagine that natural selection's yardstick was discourse itself: from elementary forms of communication (at the level of zoosemiotics) more complex communicative forms developed, which then selected more complex neural structures, which then made possible more complex and selective forms of communication….and so on and so forth. In sum, other species evolved from things while man, as a linguistic being, evolved from communication itself, that is to say from the relations with other human beings. Man is a "kind of communication". He is specialized for this function. (Prodi 1987b: 154)

"Categorial logic" (following to the "material logic" of the world) or the logic of the living is a logic based on complementarity and assimilation, where the object is either directly internalized or completely ignored. The constitution of ever more complex organisms, able to "read" the environment at different depths, allows the emergence of a different and specifically human kind of logic—"propositional" logic. What Prodi generically refers to as the "reader" (e.g. the RNA with respect to the codon of a DNA molecule, a virus with respect to the surface of a cell, a bee with respect to a flower, or a vervet monkey with respect to an eagle, a river with respect to its riverbed) can know its environment even when it does not have an immediate objective to realize within it and can know it even if no action is engendered by this knowledge. This possibility presupposes the existence, in the reader, of a (categorial) memory (akin to Pierce's "habit") and therefore the capacity of internally reproducing models of external objects. Human semiosis is primarily directed towards internal referents, more so than towards external objects.

> [w]hen the evolution of the reader allows it, the internal recording of analogy can accompany a categorial reading, applied to the outside. This internal recording would then allow the availability of the analogy beyond the event of categorial readings, capable of "standing for" such an event. With the construction of artificial analogies it becomes possible to move from a merely consuming knowledge-selection — where categorial recognition is simply an occasion for the metabolism, helping the reader's survival — to a non-destructive categorial condition of storing, and consequent availability, of "internal signs", i.e. analogies for or signs-referents of […] further levels of reading for the reader. […] The historical

reader, necessarily irreducible to the punctual attrition of its surroundings, is this constituted. This availability of stored analogies (artificial analogies insofar "non-destructive artefacts") is nothing but memory. (Prodi 1982: 91)

"Propositional" logic, the logic inherent to human language, begins with this distancing from the object, now identified as such and not merely in relation to a reader
who categorizes it: "the dawn of knowledge is therefore a *moral* event — a fuga-
cious priority is assigned to that which stands outside" (Prodi 1987b: 71; cf. Chap.
8). The virus, for example, "reads" the surface of a cell in order to find a way in and
thus that surface, as Prodi wrote, is nothing but "an occasion for the metabolism,
helping the reader's survival". So, the cellular surface is relevant to the virus only
insofar as it can facilitate its survival. In this sense, there is nothing "moral" in this
act of "reading": the use of the referent thing is an immediate consumption and
metabolizing. Conversely, in human semiosis a "non-destructive categorial condition" becomes possible: it is now possible to refer to the object even when this is not
immediately needed. Memory is this non-destructive capacity. And it is a "moral"
capacity because it preserves the object; it does not immediately metabolize it.

More specifically, the development of propositional logic—the kind of logic
needed for abstract mental operations—requires two conditions: (1) the presence of
a memory that would allow the preservation of internal traces of signs and (2) the
systematic interaction between these internal traces. From this point of view, the
very *possibility* of a syntax is subordinated to the existence of a critical mass of
signs, composing a "closed system" (Prodi 1982: 114). This also means that once a
critical quantitative threshold is crossed,[4] language becomes not simply a more or
less complex mirroring of external objects and/or of internal expressive needs (as it
is the case for the communication of non-human animals) but becomes an autonomous system, whose internal organization depends on the signs that compose it.
That is, the semiotic mass gets internally organized in a way that is not anymore
directly dictated by the external environment. In this sense, syntax and logic are
linked together; they were born together:

> [h]uman logic arises when it becomes possible to use unitarily this whole reservoir of data
> (internal signs, analogies, etc). That is to say when the subject, for a brief or for a longer
> period of time, can say of this internal complex — which can be summoned at will — *this
> is my whole*. The terms "whole" and "universal" have no objective meaning. What is the
> "whole"? What does it mean? Instead, it is more meaningful to say "this is the whole", or
> "I consider this complex system in a unitary manner". Whole and one are the same thing,
> and constitute an operation or a set of operations.
>   Starting from the internal signs — surely present in all animals — complexity can
> account for the emergence, out of a coordinate assemblage of compresent elements, of the
> capacity of conducting logical operations on the whole (to divide it into parts, to disjunct it,

---

[4] It is interesting to note that the number of signs included in non-human animal's communications
systems—natural or artificial (i.e. taught by humans)—seems to be vastly smaller as compared to
human languages. It is possible that a reduced number of signs make it impossible to form a system
and therefore sign-to-sign associations (Peppergerg 2017). In fact, the majority of linguistic signs
refers to other signs, rather than external objects. Perhaps a paradoxical Prodian definition of language could be this one: language is a semiotic device the main function of which is to refer to language itself. This is but another application of the model of the circle to a biological phenomenon.

to negate it, to sum it up, and so on — all the operations of formal logic). This is the beginning of human specificity. Discourse itself rests upon this logic. The fact that this reservoir is ordered entails that there is no such logic without a semiotic (since this reservoir will be characterized by having each of its elements distinct from the others and yet in relation with each other). On the other hand, semiotics itself would be unconceivable without a working logic (a logic without a syntax). (Prodi 1987b: 152)

It should be repeated that in the course of this process, the subject—that Cartesian semiotics sees as an uncaused cause—is constructed *as* a subject together with the natural relations that both precede it and make it possible (I will explain this process in detail in the next chapter). Let me go through the steps we have taken so far (something Prodi often does in his own books). "Categorial logic", the world of life, is the space of selections, that is to say the space of interactions (or non-interactions) between single organisms and objects in their environment. At the beginning there is a certain thing (A) entering into a relation with another thing (B) while ignoring a third thing (C). From this interaction a more complex system is created—AB—for which further and unexpected selective activities now become possible. A is not a Cartesian subject: it is a thing, a reader, which can selectively interact with "its" environment (actually, following Jakob von Uexküll, we should say that the environment itself is nothing but this selection) thanks to both its own composition and that of other things that surround it. As these selection operations become more complex, a reader composed of a multiplicity of subordinated readers spontaneously constitutes itself: indeed a reader is a "federation of internal readings, granting advantages to the global reader, and constituting it as a stable interpreter of its reality" (Prodi 1987b: 141). The resulting whole—the reader—does not precede the existence of its parts nor can it boast some kind of autonomy with regard to them. Indeed, "it would be utterly misleading to think of higher structures as monads capable of eating/reading those at a lower level. The higher structure does not miraculously emerge out of nothing, but raises from simpler interpretive structures, and is therefore always, in itself, a federation of readings adapted to the signs that are read" (Prodi 1977: 83).

In the framework of Prodi's semiotic anthropology, the human subject becomes the last manifestation and articulation of the world's primordial material logic. The subject is born from the world; or, more precisely, the subject is the world coming to know itself through one of its parts. The connection between the subject and the world is possible precisely because the subject is a fragment of the world; the subject is nothing but a transformation of the object.

[t]he philosophical problem of how the mind can know the world must necessarily *be articulated through the elementary roots of our biological knowledge.* […] In other words, there is a logic of relations between non-selective things […]: material logic. In a restricted domain this organizes itself and becomes selective logic […] [that is] categorial logic, or bio-logic. This, in turn, complicates itself and thus gives rise to logic in the standard use of the term, human logic […], propositional logic. Every new form of logic is included in the one that precedes it, it represents a more limited and sophisticated domain than the previous, and exploits the possibilities afforded by it. So, there is no categorial logic outside of the energy exchanges and structures made possible by the conditions of material logic, and there is no human (propositional) logic outside of the selectivity conditions instituted by categorial logic (bio-logic). (Prodi 1987b: 145)

Chapter 6
From Complementarity to Semiosis

In our view, it is not symbolic language which "explains" the machinery of the cell, but it is this machinery that must explain (through a suitable complexity) the nature of a symbolic language.

(Prodi 2010: 334)

Abstract Giorgio Prodi poses a fundamental philosophical problem: how was the emergence of immaterial meaning possible within the world of material things? In order to answer this question, Prodi develops an original relational ontology. The world is made of relations between things, not of things. The world is relation; semiosis is relation: it follows that the world itself is semiosis. This chapter will expose Prodi's natural history of meaning.

Keywords Complementarity · Selection · Proto-semiosis · Phylogenesis of semiosis

According to a classical schema, the fundamental semiotic relationship can be represented, as a first approximation, by a triangle (as shown in Fig. 6.1) where the sign is in a relation with a given object—its referent—through the mediation of a meaning (Eco 1973: 25).

Within Prodi's project of a biological foundation of semiosis, this schema presents a problem (one we have already examined, even if in a different form): what is this "meaning" that mediates between the signifier and its referent? What is its biological origin? The most common answers given to these questions are (1) meaning depends on the subject's intention, i.e. "meaning" is some unspecified "mental" entity, or (2) it depends on a convention established by two or more subjects, which in turn becomes the institution of a code. In both cases there is a direct line of correspondence between meanings and their referents and an indirect line between the signifier and the referent. Following Prodi, we have seen how the first answer is completely unwarranted, since it is based on an utterly mysterious principle—the intention, or consciousness, of the subject—while a good explanation should always

Fig. 6.1 The semiotic
triangle according to Eco

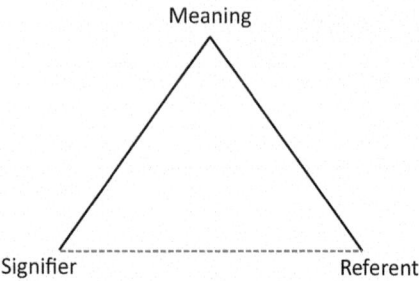

Meaning

Signifier   Referent

be grounded on simpler facts than those it purports to explain. The second answer—
that pointing to a conventional agreement between subjects—is also unsatisfactory,
since it does nothing but extend the first (pseudo-)explanation to the entire community bound by a given set of norms: if there is no real justification for the fact that for
a subject the sign S is in relation to object O, why would we feel reassured by the fact
that many subjects agree to such a convention? If a relation is unjustified for one
subject, it will be even more unjustified if we multiply the subjects who agree to it.
But there is also another possible solution, based on the concept of iconicity.
According to this hypothesis, the sign would be in relation with its referent because
it would somehow resemble it. In reality, this is not a viable solution, since the relation of resemblance is not a "natural" relation, but it is grounded on a previous—even
if unconscious—institution of such resemblance. That is to say, if a thing Q resem-
bles another thing Q_1 that is not because of a natural resemblance, but because someone detects in Q_1 a resemblance with Q; there is always someone who adjudicates
this resemblance, which is not something given in nature but will always lay in the
eyes of the beholder. Following Eco's (1976) classic critique of the notion of iconicity, we can say that the latter presupposes a rule (a rule that can be an unconscious
rule but a rule nonetheless) and that, therefore, it would establish a somewhat arbitrary relation. Compared to Eco's position, though, Prodi's own stance features a
rather decisive difference: for Eco this critique was meant to underline the radically
arbitrary (and therefore subjective) character of semiosis; Prodi, on the other hand,
exploits it to pave the way towards a biological model of semiosis, a model that is not
grounded on either arbitrariness nor iconicity. In fact, the two possible explanations
we have just considered are both examples of linear explanations (the first one top-
down, while the second bottom-up) which are constitutively unfit to account for biological phenomena, i.e. phenomena based on the notion of continuity.

A way out of this impasse, albeit a paradoxical one, is that chosen by structuralist
semiotics, resolving the issue by presupposing that, actually, the problem of the link
between signifier and signified "simply" is not a problem at all. Indeed, Eco argues
that the problem of the link between sign and thing is a false problem (the so-called
referential fallacy [Eco 1976: 58]).[1] According to this radical stance, the concept of

---

[1] In later years (see particularly Eco 1999, Chap. 2), Eco will relax his critique to the notion of
"referent", although without fully changing his mind regarding the concepts of iconicity and
semiosis.

referent would be superfluous for a complete theory of semiosis. For Prodi's biology-informed worldview, this is clearly an unsatisfactory solution, since it would have the entire semiosic system rest upon nothing at all or—in a very similar formulation—on itself. Just like the Cartesian semiotic subject appears to be a mystery, so the entire process of (human and non-human) semiosis becomes thoroughly mysterious and unmoored from the material world of things. At the same time such a solution does not answer the genetic question: what is the natural history of semiosis? Once again, Prodi is faced with the spectre of dualism. Once the subject is placed in an autonomous and independent position, semiosis becomes opaque to explanations: in both its non-human and human forms (Cimatti 1998; Martinelli and Lehto 2009; Witzany 2014), as well as within the plant kingdom (Kull 2000). In Prodi's perspective, on the other hand, the relationship between the reader and thing-read is the (proto-semiosic) starting point of the whole of natural semiosis.

> [i]t is certain that the referent has a complex relation with human consciousness, and with cultural semiotic processes. For this reason, it is often kept quarantined: far from discourse, like a kind of scary boogeymen. The semiotic methods for coming in contact with the environment are extremely complex, and it seems childish to have things directly intervene as protagonists of our act of naming them. But if the things themselves can be modified, in a precise and predictable way, by scientific consciousness (that is to say by a particular way of naming them) this means that some link with the referent must exist somewhere in the chain of signification. It is not advisable to adopt an idealist strategy in order to get rid of the referential fallacy. Solidarity with the referent must be sought elsewhere, in two ways: first by making ourselves aware of the complexity of human knowledge, through a critical and informed epistemology (looking at how science classifies objects) and, secondly, by taking the birth of the function of exploration as an object of study. This second path shows that, initially, the referent is not at all "fallacious", in fact it coincides, with no need for mediations or rules, with the sign. Obviously, it is a "sign for" something capable of "being specifically modified by" it. (Prodi 1977: 32–33)

Here the reference to Eco's "referential fallacy" is explicit: we should not be scared of it. On the contrary, only by giving back to the thing—to the referent—its role in semiosis it will be then possible to assign it a naturalist foundation. At the beginning the sign coincides with the thing, because the thing complements the reader, and vice versa. Everything selectively adapts itself to the things that surround it, "choosing" some and "discarding" others: "chosen" things are those with which a link is possible, in order to form a more complex assemblage. The "chosen" thing is the one that "means" a link and therefore survival. In order to answer the question I posed at the beginning of this chapter—what is the foundation and the origin of the semiosic relation?—it is therefore necessary to discard the semiotic triangle and rather ask ourselves how the first proto-semiotic interactions are born in *nature* (i.e. in the context of life phenomena). Prodi's starting point is the notion of natural *meaningfulness*: "the more general condition of a language situation (the search and transmission of sense) is meaning, that is the condition for which a natural presence is correlated to another natural presence through a relation of selectivity. This condition is not only at the foundation of language as commonly intended but also grounds the whole of biological organization. From this point of view, the latter is intrinsically linguistic" (Prodi 1983b: 186).

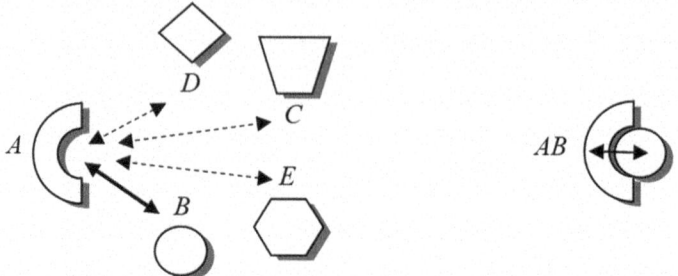

Fig. 6.2 Natural semiosis as natural meaningfulness

Let us look at Prodi's model in greater detail, starting with an extremely elementary scenario (illustrated in Fig. 6.2): an organism-thing *A* that "explores" its environment. To say that *A* "explores" means that it "freely" moves within the environment—there is no need to presuppose any conscious will driving this behaviour. To offer a simplified model, we can postulate that in *A*'s environment, there are only four things: *B, C, D,* and *E. A* is a particular material structure: for example, if it was a protein, the atoms composing it would give it a particular steric configuration, that is to say a specific capacity to interface with determinate atomic configurations, excluding many others. In *A*'s environment, the only thing that *A* can interface with is *B*; that is to say between *A* and *B*, there is a relation of specific *complementarity*. Employing a metaphor often used by Prodi, between *A* and *B* there is a relation akin to that between a lock and the unique (at least in theory) key that can open it. With respect to *A, B* is the only *meaningful* (or, in other terms, *pertinent*) portion of its environment: "meaning emerges as categorizing: it is a sign, that is to say selective interaction" (Prodi 1982: 169). To *A, B* represents a sign that "stands for" the relation that can be established with *A*. However, at this rather basic semiosic level, *B* as a sign is identical to its referent, that is to say *B* does not "send to" anything different than itself: in *B*, sign and referent coincide. It is therefore a "direct and biunivocal thing-reader relation (where sign and referent coincide)" (Prodi 1982: 169). Properly speaking the semiotic triangle illustrated in Fig. 5.1 is a far (and essentially unreachable) goal, since "initially the terms of the semiotic relationship are just two [...] mediation (the autonomy of the "sign") is a later development" (Prodi 1977: 158). To *A, B* means something to interface with, and with whom to form a more complex system, the assemblage *AB*:

> an object *A*, casually coming in contact with *n* surrounding objects, does not enter into relation (does not react) with any of them except one, the object *B*. The *n* other objects are indifferent to *A*, while *B* is meaningful. When the specific reaction is, for example, the construction of the assemblage *AB, A*'s exploration of its surroundings entails a 'complication', that is to say the appearance of the stable assemblage *AB*, previously non-existent. (Prodi 1983b: 186)

At first, there is only spatial congruence: possible in one case, impossible in all others. Nothing more. "Meaning" means possible relation: there is no intentionality, consciousness, or will in this natural meaning. It is natural because it is not arbitrary. *A* establishes a relation with *B*, forming the assemblage *AB*, simply because *A* is thus and so, and this structure can be adapted to *B*'s and vice versa: "if we want to represent the proto-semiotic sign in 'triangular' manner, on one vertex we will place the thing, on another the reader, and on the last one the history of the relations between the two, that is to say the genetic 'depth' of their relation of specificity" (Prodi 1982: 169). In fact, *A* and B have progressively adapted to each other, (this is the "history of the relations between the two"). The clearest example of this natural complementarity is the so-called genetic code, an example often used by Prodi. In reality this is not at all a code, since codes are arbitrary: the genetic code, on the other hand, is neither arbitrary nor conventional:

> [w]hat is the mechanism that links the invariant sequence of [DNA's] bases with the structure of the resulting proteins? There is a correspondence law, a law that is not established by decree, but by its being produced by nature (this, and not others). [...] Therefore, there is a correspondence between the sequence of DNA's bases and the sequence of amino acids of the respective proteins. [...] The correspondence rules between DNA and proteins, along with the mediations that occur at the nuclear and cytoplasmic level, represent the most conspicuous and general example of the kind of historical reading of meaning we labelled natural-semiotic, upon which the whole of biology is grounded. (Prodi 1989: 36–37)

Let us attempt to represent the general structure of this situation in Fig. 6.2. We should remember that the core idea is that of accounting—in a naturalist and non-Cartesian manner—for the appearance of meaning in the natural world. Here meaning means, at a first approximation, "selection" and "choice", although three is no agent who selects nor chooses.

At the beginning—a purely hypothetical moment, since to take seriously the concept of continuity implies that the beginning has always already begun—there are only some things: *A*, *B*, *C*, *D*, and *E*. *A* has a certain spatiotemporal structure—which it has not chosen, since *A* is nothing but this structure—which allows it to interlock with *B*, but not with *C*, *D*, or *E*. The same applies to *B*, whose structure is more adaptable to *A*'s, and only to *A*'s. So *A*—without knowing nor willing it—"chooses" *B*, the only thing in its environment with which it can form a larger assemblage. Of course, the same applies to *B* with respect to *A*. As Fig. 6.2 clearly shows, it is impossible to differentiate a subject from an object. *A* "chooses" the character of *B* that, from its point of view, is pertinent—*B*'s spatiotemporal structure. All other characteristics are not pertinent. This schema likely has two sources of inspiration. These are Jakob von Uexküll's functional schema and Peirce's distinction between "dynamical object" and "immediate object": "[w]e must distinguish between the Immediate Object, — i.e. the Object as represented in the sign, — and the [...] Dynamical Object, which, from the nature of things, the Sign cannot express" (CP: 8.314). The "dynamical object" is the thing, with all its infinite characteristics. Or at least the thing in itself, with no contact with anything else—if that could exist, since we must say that it really does not exist once we acknowledge that it is just a cluster of relations. The "immediate object", on the other hand, is the thing as considered from another thing: it is *B* from *A*'s point of view and vice versa.

For A, B is a sign; the meaning of B coincides with B itself. B is meaningful for A because, from A's perspective, it is a pertinent thing. In Prodi's terminology, A is the "reader" of B (but we have seen that this relation can also be considered from the other direction, such that B would be the "reader" of A). The sign is not immediately the thing, but the thing in relation to a reader: the sign is the existence of a deciphering or a relation of specificity. Thus, the problem of a foundation—of justifying the union of a sign with its referent—is completely bypassed, since B is at one and the same time sign and referent. The biological sign emerges as complementarity and therefore as continuity: "our problem is: how can this meaningful situation, where a thing is a sign for an interpreter that holds the interpretive codex, emerge in nature? [...] In this case, a sign does not have two faces, there is no *semainon* and *semainomenon*. The sign is a thing, and a thing becomes a sign the moment that a reader that can read it — i.e. that can selectively "take it up" — appears. The selection, that is to say the specific relation to which an interpreter participates, is the beginning of semiotics" (Prodi 1987b: 146).

Let me sum up the fundamental characteristics of "biology" as "natural semiotics" (Prodi 1987b: 147).

i) A's actions are purely casual, i.e. non-intentional. "Natural semiotics", unlike Cartesian semiotics, does not presuppose any signifying intention as a starting point for the semiosic process. Prodi's model, like Peirce's but more radically still, is profoundly anti-subjectivist. In Prodi's words, A performs a "'reading' [that is] a process that brings A to casually collide with its surroundings, identifying the meaningful object B, and reacting with that, and only with that. This process is A's reading of its environment" (Prodi 1983b: 186).

ii) A reacts to B, that is to say it is so structured as to be able to interlock with B. Natural meaningfulness presupposes a natural mechanics. We need to interpret this mechanics in a literal sense—a certain material form is suited to welcome some things but not others: "the enzyme is a complex molecule that, in its spatial development, has a dip, cavity, or hole that can be 'spatially' adapted to *only one type* of molecule, in order to react with it. Thus, meaning is born as biology. Meaning is identified with biology" (Prodi 1987b: 133).

iii) A's reaction with the external world is *selective*. A's action is implicitly (considering its random nature) guided by its "search" for pertinent traits: A "searches" its "environs" [*circostanti*] for what is interesting to it. When A "finds" such an "interesting" thing (in Fig. 6.2 this is represented by B), it reacts with it, forming a new assemblage. A relation is thus established between "adaptation (to be made for something else) and meaning. An organism has external correlates towards which it is adapted, and that can therefore be exploited by it. For it, they are meaningful" (Prodi 1979: 187). The relationship established between "surroundings"—i.e. A's world before a "choice" is made regarding what is pertinent and what is not—and "environment" is wholly analogous to Peirce's distinction between "immediate object" and "dynamical object":

[i]n philosophical terms, an enzyme is a reader that "categorizes" reality determining the set of all the molecules it can factually react with. The term category is used here both in a Kantian sense (the "point of view" of a reader who gives order to the world) and an Aristotelian sense (the objective ensemble of things presenting a given character). This semiotics (or proto-semiotics) is the basic feature of the whole biological organization (protein synthesis, metabolism, hormonal activity, transmission of nervous impulses, and so on). In all these cases a specific relationship between a reader and its sign is clearly established. (Prodi 2010: 329)

iv) Through the complementarity relation that is established between them, A and B come to form the assemblage AB. In turn, this new thing in the world—AB—"searches" for other meaningful things in the world and so on. This model accounts for the possible creation of ever more complex readers, those that, as we have seen, Prodi defines a "federation of readers". This is an important possibility because, in perspective, it will explain the possibility for a sign to eventually "detach" itself from its referent. The passage from A and B to AB is a theoretically crucial point. The situation described in Fig. 6.2 is extremely simplistic and rudimental, and it is very difficult to imagine how, even in line of principle, it could apply to far more complex situations and in particular to human language. Prodi is always careful to preserve the principle of continuity, and therefore he has no other choice but admitting that the complex is nothing but a quantitative complication of the simple. He finds the solution of this problem—*the* problem of his entire philosophical-scientific project—in the construction of ever more articulated and internally differentiated readers:

[t]he phylogenetic constitution of an individual around one or many objects that are meaningful for it (as well as external environmental correlations, which too become meaningful) implies a very long series of selective choices in the past of progenitors, and therefore a codified storage of order: thus are formed the "categories" of knowledge, i.e. the physical sections of exchanges between reader and things read — the modalities of exchange — that are the criteria for meaningfulness (not in an abstract sense, but relative to the organism for whom that set of external things results meaningful). The evolution of consciousness — of which human consciousness is its most complex form — must be interpreted according to this model. The specialization of communicative functions implies an enlargement of the area of meaning: a very differentiated individual is capable of exploiting/knowing a very wide area of its environment. The environmental area of a bacteria is restricted, since it can only exploit/know (recognize as meaningful) a very narrow range of environmental elements—for example nitrates or glucose. This area is enlarged following the appearance, in the environment, of more complex beings capable of moving within it and "reading" many more things, including other beings — similar to or different from them. Further, this area is massively enlarged in the case of man, with the appearance of strong functions for hypothetical simulation, together with the capacity of storing and transmitting the data of experience. (Prodi 1979: 188–189)

In Fig. 6.3 I try to further simplify the situation portrayed in Fig. 6.2. Figure 6.3 shows how the semiotic triangle (Fig. 6.1) can actually be flattened onto a biunivocal and complementary relation between reader-thing A and read-thing B (the sign/referent). From this relation will then derive the formation of the complex thing AB. This simplification is expedient in order to demonstrate how Thirdness can be

Fig. 6.3   The flattened triangle of biosemiosis

reduced to Secondness. Prodi's entire project is grounded on this reduction. Either semiosis is reducible to an interaction between things (and therefore, in Peirce's terms, to Secondness) or it will forever remain inexplicable from a biological and naturalistic point of view. Either semiosis is a biosemiosis or it is a form of idealism.

The most important biological characteristic in this diagram is that the entire process is thoroughly non-intentional. Meaning arises, literally, *from things*, and it is a transformation of things:

> given an assemblage *AB*, this is composed by *A*'s reading of *B*, and by *B*'s reading of *A*. In an elementary condition, no one thing plays a larger reading role in its environment, but everything reads and is read in the same way, being at the same time reader and part of the surroundings. That is to say, a world of predeterminate readers does not exist, and reading is a natural process of research of reciprocally significative conditions. (Prodi 1983b: 180)

Until the formation a complex thing *AB*, the natural reading process between *A* and *B* goes both ways. There are no hierarchies in nature; there are only things and relations between things. Therefore, *A*'s act of reading *B* can be also interpreted in the other direction, from *B* towards *A*: this explains why, in Fig. 6.3, there is also a dotted arrow going from *B* to *A*. Obviously, in this case *A* would become a sign/referent for *B*. Once the assemblage *AB* is composed though, the process of "selection" and "choice" begins anew, and a new "environment" is extracted from a "surroundings" of things:

> [s]o *A* selectively reacts to *B* alone. This *A* "takes into consideration" external reality (i.e. other *A*s, *B*s, *C*s, *D*s…) because it can collide with everything else, in a purely statistical and random kinetic situation. All other permitted states (*C*, *D*, and so on) are indifferent to it, while *B* is meaningful. I say that it is considered factually meaningful simply because it selectively reacts with *B* and not with anything else. Meaning is what we detect: it is the assemblage that was created (without implying, with this term, anything intentional—we need to eliminate any anthropomorphic overtone from this model). […] We say that *A* reads reality and finds in it *its* meaning, that is *B*. […] *A knows reality through its selective reaction with B*. everything else is indifferent, only *B* is meaningful (and vice versa). *A* knows reality because it reacts with B. *B is "its" sign*. (Prodi 1987b: 131–132)

It is necessary to reiterate an important point, something that often is forgotten but that is crucial to understand Prodi's model. The thing, for Prodi, *coincides* with the relations that it establishes—or can establish—with the other things it encounters: "the starting point must be sought in the event, rather than in 'things'" (Prodi 1982: 21). Let us return to the thing-reader *A*; this thing is individuated as an isolated and distinct reality by the process it partakes in only for the purpose of analytical clarity and exposition. In reality, *A* is nothing but the *temporary* outcome of a previous biological relation, destined to be transformed into new and more complex relations, and so on without an end. Essentially, *A* is only the waypoint of an endless process of semiosis, since if such process would end, this would entail the end of life: without life, there is no semiosis. The thing is therefore always an event:

> we could say that at the foundation of material logic [...] there is 'change' if only this wouldn't always lead us to interpret this as the change of an already-given *thing*. Existence self-maintains itself as a networking of events, that is to say an operative factuality of logical relations [material logic]. (Prodi 1982: 27)

The reader as an entity can only be individuated *after* a relation between it and some other meaningful thing has been established; the intention of establishing a semiotic relation is the result—and not the starting point—of the relation itself: "the reader constitutes itself with environmental terms and *through* environmental terms. There is no reader that precedes the mode of its formation, and the latter is the selective relation that grants it advantages" (Prodi 1983b: 188).

Upon the foundation of proto-semiotic assemblages such as *AB* (in Fig. 6.3) more extended chains of natural meaning are then formed, i.e. living organisms. These are precisely "complex federations of meaning/reading, which are coherent and capable of expressing their newly achieved reading advantage onto their surroundings. Herein lies the intrinsically 'linguistic' character of biological objects" (Prodi 1983b: 189). How should we interpret the adjective "linguistic"? It should be understood as expressing the fact that life phenomena are intrinsically relational and not in the sense that they would be literally made of language. On the contrary, language is the most complex form of semiosis. The natural world is the world of semiosis:

> [p]ropositional logic, therefore, presupposes and extends categoriality. It presupposes the concrete fact of existence of the organisms that reached categoriality, and precisely its advanced and complex forms, since categoriality is itself a nexus of relations of material logic, including the latter within itself. Propositional logic too is therefore firmly in contact with its objective reading terms. (Prodi 1982: 84)

Following these methodological guidelines, we can start to understand what, for Prodi, a generic "mind" can be: "more of a network (in its decentralized globality) than [...] a unity of a hierarchical kind, representable as a system of filters arranged in a vertical manner" (Prodi 1983b: 176). An example of a hierarchical system is the classical Cartesian model, placing consciousness at the top of the mental apparatus, as an uncaused cause, a wholly different substance from the rest of the body. This kind of notion is incompatible with Prodi's naturalistic model: "we are not trying to explain human language through an extra-natural intervention of 'intentional' kind" (Prodi 1983b: 314). For Prodi the mind is a "network", a complex federation of chains of meaning-reading devices. With an intuition still very timely today, Prodi understands the complexity of the mental not as intentionality—intended as a primordial and irreducible characteristics of the mind (Chalmers 1996)—but rather as a property emerging from the non-intentional interaction of an enormous number of extremely simple structures, in themselves utterly unintelligent: "to explain the mind, we have to show how minds are built from mindless stuff, from parts that are much smaller and simpler than anything we'd consider smart. Unless we can explain the mind in terms of things that have no thoughts or feelings of their own, we'll only have gone in circle" (Minsky 1985: 18).

We have seen in Fig. 6.2 how the primordial proto-semiotic relation occurs. It remains now to analyse how it is possible to subsequently move towards a situation where a sign and a referent get detached to the point of *looking like* separate entities—a situation conducing to a sign properly speaking, the sign of traditional semiotic triangle that "refers to" something that can be physically *absent*. In fact, a situation the outcome of which is the classical semiotic triangle. Prodi's proposal can be considered valid only insofar as it can explain this transformation. It is necessary to give an account of semiosis, and in particular of human semiosis, starting from the "extremely simple" biosemiotic complementarity. The problem is that, while natural meaningfulness is always based on presence—the reader directly touches the thing it reads—in the case of a sign as a "sending to", meaning is not materially present. In this case semiosis (again, in particular human semiosis) is in absentia, while biosemiosis is always *in presentia*.

This process is described by Prodi as an evolution, developing in three steps, the first of which is exemplified in the situation described in Fig. 6.2. The precondition of this evolution is the development of ever more complex readers, establishing relations with their environment according to ever more complex modalities: "complication, as a natural fact, is based on the advantages that this endows to structures" (Prodi 1977: 63). Organism *A* in Fig. 6.2 holds a relation of immediate complementarity with its sign-referent *B*. It merges with it, it physically assimilates it. That is to say that *A* is not capable of interacting with *B* in any other way. The first step for the development of more elaborate forms of semiotic relation, Prodi argues, is the presence of more complex readers. In order for this to happen, it is necessary that the situation illustrated in Fig. 6.2 be repeated several times, in order to form more complex and unitary "federations of readers". The assemblage *AB*, for example, will now be able to "read" more objects of its environment, which were beyond *A* and *B*'s ability of categorization. Even more extended organisms will now be able to form, and so on all the way to the emergence of complex systems describable as "federations of readers" or—using another of Prodi's definitions—as a "codex [...] a categorial bundle, that is to say a group of reactive models and of recognitional capacities" (Prodi 1977: 63).

This also means that the "federations of readers" are intrinsically semiotic and that the reader is in fact made of semiosis. The most direct inspiration for this radical thesis of Prodi's is once again Peirce who, in "Some Consequences of Four Incapacities" (1868), writes quite explicitly that:

> there is no element whatever of man's consciousness which has not something corresponding to it in the word; and the reason is obvious. It is that the word or sign which man uses is the man himself. For, as the fact that every thought is a sign, taken in conjunction with the fact that life is a train of thought, proves that man is a sign; so, that every thought is an external sign, proves that man is an external sign. That is to say, the man and the external sign are identical, in the same sense in which the words homo and man are identical. Thus my language is the sum total of myself; for the man is the thought. (CP: 5.314)

Prodi's other point of reference is again Wittgenstein who, in the *Tractatus*, writes "[t]he limits of my language mean the limits of my world" (Prop. 5.6) (in Chaps. 9 and 10, I shall analyse the ethical and aesthetical consequences of this thesis). The human "mind", for Prodi, is not a *substance* (and here he is really quite close to Peirce), but rather it is a long and unending process or "translation chain" (Prodi 1983b: 190) of operations of categorization/identification that give rise to ever more complex and articulate semiosic assemblages.

Such a system can (1) behave as a unitary whole yet being nothing more than a "federal" constellation of parts without a centre and (2) categorize reality in more sophisticated ways. For example, the process of recognition of a sign/referent by a complex reader does not necessarily terminate with the latter assimilation of the former—as in the case of the relation between A and B. Let us imagine a complex reader N, facing object O: the "simple" perception of O already implies a selection among possible reactions that N could have with O. While A could only perform *one* action—to recognize and assimilate B—a multiplicity of alternative "translation chains" is now active within N (here too the Peirce distinction between "immediate object" and "dynamical object" applies). To a higher complexity of the reader corresponds a higher complexity of possible responses and a wider range of actions. System A can interact with its environment along a single dimension—merging with B—while N, as a "categorial bundle" or a "federation of readers", can interact with reality in different ways, being itself a complex structure with an internal stratification of a rich network of "translation chains". In particular, this means that when N encounters the sign/referent O: "the encounter with the sign is [...] preserved, and parallel to the process of dissipation there remains a stored component.[...] An internal analogic situation is thus established: an internal sign. This sign 'endures', i.e. it is available to the structure for a longer period than the act of interpretation that produced it. This sign, precisely due to its being analogical and complementary, has nothing to do with the nature of the factual sign from which it derives [...]: it lives into, and it is contained by, the structure, while the sign and the interpretation can be external, exhausted, and lost" (Prodi 1977: 118–119).

The decisive step in the development of ever more articulate forms of semiosis, farther from the material encounter with the thing-read, is when traces of the interactions with the environment are formed inside the reader. The turning point, that is, is the formation of an internal memory. The living world cannot survive without some form of memory (Honig and James 1971; Braitenberg 1984; Roitblat 2014). The necessary condition for the formation of these traces is that the organism be a complex structure, where the relation between A and B would no longer be direct but rather pass through a multiplicity of intermediate translations. At a certain point it can be hypothesized that the presence of B determines permanent changes to the internal structure of A, traces that can be reactivated when B will no longer be present. The transition to ever more complex forms of semiosis, where sign and referent are distinct, implies therefore the existence of *memory*. Through memory it is possible to fixate associations that allow for the existence of a *codex*, that is a sequence of sign-object pairings. But in order for a codex to emerge, a community of systems that read the world analogously is also necessary; a certain animal call is a sign for a given species of animals, while it is a "simple" noise, or at best a clue about the presence of a living being, for another species:

> [a]t this stage there are at least two types of sign. One is the reader-nature sign, the other the reader-reader sign. An example of the first kind of sign is a trace (a scent) that guides towards food and that "stands for" (directly or indirectly) the food. [...] An example of the second type is a mating call. It is an inferential symptom of a linguistic state, albeit imprinted and elementary. These are two rather similar conditions, but the second type of sign already belongs to a discursive society. In both cases, the arrival of a physical signal finds a response: at this level a signal is not characterized by its being perceivable — something that belongs to the previous stage, to the organization of proto-semantic translation chains. [...] Rather, what characterizes this stage is a further selectivity, standing on the shoulders of the previous stage. An animal can hear sounds, and this belongs to elementary semiotics [...] This is categorial. [...] But once we presuppose this condition, a sign of a higher level raises above because, among a series of noises indifferently perceived the animal selects (i.e. interprets categorially) the one sound that [...] is for meaningful for it. [...] Categoriality or specificity [...] has in this circumstance to do with an ever more defined associative life, with some characteristic behaviours and, ultimately, with a history: yet a more evolved, more recent, more group-specific part of history. (Prodi 1977: 170–171)

The first dissociation between sign and referent requires the presence of complex systems, endowed with memory and capable of forming, and maintaining, some primordial forms of a codex. This therefore implies a first foray into the world of cultural history, inscribed into the genome. In other words, it implies a first form of learning. An example, in nature, of this phenomenon is what ethologists call a process of *ritualization*, through which a non-semiosic functional animal behaviour is transformed into a sign for something else—for example, the intention to attack a rival (Lorenz 1981). This process means that, at first, a certain behaviour is functional for a certain purpose—for example, a wolf baring its fangs as a preparation for the attack itself. Let us imagine, then, the following predicament: wolf *W* is about to attack wolf *P*. *W* bears its fangs (an action justified by his intention of biting the other wolf), but before the attack, *P*—capable of perceiving typical behaviours of the species to which it belongs—infers the imminence of an attack from the baring of *W*'s teeth, and therefore retreats before this can actually take place (similarly, it could also be the case that *P* had previously observed the sequence "baring of fangs → aggression" and has therefore concluded that there is a correlation between the two events. That is to say, *P* "understands" that if the first event takes place, the second will follow or that the first makes it possible to predict the second—this is an example of "categorial logic"). For *P*, *W*'s (or any other wolf's) baring of fangs has become a sign that "sends to" a possible aggression. In time, *P*'s ability to infer *W*'s intentions from its behaviour—an evolutionary advantageous one (it is easy to imagine how an animal gifted with *P*'s inferential ability would survive longer than one that isn't)—spreads in the group, meaning that the genetic make-up that makes this ability possible in new wolves spreads in the wolf genome over those genes that do not allow for it (and thanks to this capacity *P*'s fitness is greater than wolves who do not have it). As time goes on, *P*'s inferential ability becomes common in the genetic pool of the entire group, and this makes it possible for the functional act of baring the fangs to become a sign for the *intention* to attack. That gesture has undergone a process of ritualization, it has *become* a sign. In turn, this entails that together with the sign, a codex has emerged, codifying the association of a certain behaviour with a meaning. However, this is an innate codex, specified by the animal's genome: P's ability to "read" W's behaviour, indeed, critically depends on its genetic heritage: "*the transcendental, and synthetic a priori judgments, are only my ancestors*" (Prodi 1979: 186).

The sign that has been established in this community does not depend on other signs; it is not bound to other semiotic capabilities and—most importantly—does not require learning. In fact, the substantial fixity of this semiotic association is confirmed by the fact that (1) it is very hard—albeit not impossible (Cimatti 1998)—to find, in nature, convincing examples of a lie.[2] To lie means to explicitly assert a falsehood, not "simply" to omit a part of the truth. In order to lie, one needs to "freely" use the codex and to be able to use a sign in the absence of the thing-referent. This seems to be an extremely rare behaviour (but not necessarily impossible, see Sebeok 1986) in the non-human animal world. Further, (2) none of these signs, in nature, is properly speaking learnt. That is to say, a new-born non-human animal is born with the capacity of using all the signs of its species' language and at best needs to learn how to use them correctly (in the right context), but it does not need to learn the signs themselves nor can it learn new ones (Fouts 1997). At this level there already are signs distinct from their referents (and therefore we are already beyond the situation exemplified in Fig. 6.2), and yet these signs do not yet have a life of their own, they do not compose a properly historical system to be learnt that could be modified and extended:

> [t]his type of sign is characterized by a continuous connection: outside-inside-outside — which does not preclude, on the inside, the occurrence of re-elaborations and interferences, but that makes autonomous operations of simulation impossible. In other words, these signs are defined as such by a genetically imprinted logic, still belonging to the categorial domain. A simulation can exist, but only as belonging to a mirrored chain. [...] From the point of view of the reading relation (rather than that of the genesis of reading systems) this is a biunivocal communication, happening without the intervention of a real linguistic corpus. Therefore, at this level, we find neither a propositional mediation of the structure of the reader, nor a linguistic mediation by the community of readers. There is a codex, but it is in the genome. (Prodi 1982: 172)

The turning point in the evolution of semiosic relations takes place when the set of the signs in the codex crosses a certain critical threshold, becoming a complete whole that can be learnt, capable of changing the structure of the reader from within. Beyond this threshold, the internal signs form connections that are independent from what can be directly imposed by the environment. The central element in this new semiosic level is the "reader-reader sign" (Prodi 1982: 170). The thing-referent of a sign is *another* reader:

> the extremely complex functions related to logic, language, and hypothesis (and to all aspects of human knowledge) are produced through complexity, that is through the emer-

---

[2] The empirical question is highly controversial, and a much hangs on just how stringent the criteria to define a behaviour as a "lie" are taken to be (see, e.g. Gómez and Martín-Andrade 2002).

gence of new combinatorial and connective possibilities of fundamentally common structures; not through qualitatively new units and structures". (Prodi 1987b: 52)

The internal signs compose a network that can (or cannot) be linked to some external entity because its existence does not depend on that of the external objects to which, as a whole, it is connected. The "translation chain" that connects the sign-referent to the internal network, that is to say to the mind (the mind is the "network"), is now enormously more complex, such that between sign and object there no longer is the "biunivocal communication" still binding together the sign to a certain behaviour exhibited at the lower semiosic level. If, at the first level, the sign still coincides with the referent, at the second level a meaning takes shape, which "sends to" a thing-referent, as illustrated in Fig. 6.4 (this is the only diagram explicitly proposed by Prodi himself [1982: 174]).

Only at this evolutionary level can we, for the first time, speak of something akin to the classic semiotic triangle, although Prodi tellingly presents it only as a linear relation rather than in its "traditional" form. Prodi profoundly dislikes the semiotic triangle:

> if a system is very complex, such that it is impossible to single out the steps of the translation chains, it is therefore necessary to speak, globally, of 'triadicity' of the sign: but the fact remains that this kind of semiotic approach is made necessary by our ignorance about the phenomenon's physiology. In itself, the sign is always a reading process of something else, that is to say it is the thing *qua* meaningful for a reader. (Prodi 1982: 97–98)

Moreover, this diagram does not have a preferential vector, and most of all, it does not begin—like it is often the case—from the sign but rather from the thing. Here we can discern the limit of the semiotic triangle: by assuming the sign as a starting point, that model unwittingly places itself in the tradition of Cartesian semiotics. From whence does the sign come, placed at the top of the semiotic triangle? To start from the thing, on the other hand, means to acknowledge that nature and the world come first, followed by a reader, and by that reader capable of "discourse", i.e. the human animal, later. In Fig. 6.4 the intermediate level of "representation" appears (really an infinitely extended level). The passage from Fig. 6.2 to Fig. 6.4 is a crucial one. For the first time in the natural history of semiosis, there is an intermediate level between reader and thing-read: the level of *meaning*. *Meaning* is neither the referent-thing nor the mnemonic trace of the perceived object—rather, it is a purely semiosic entity. The thing-referent is a thing of the world, while the mnemonic trace is an entity in the brain. We should recall that Prodi's challenge is that of accounting for the *meaning* of semiosis, and if this meaning was a thing—either in the world or in the brain—there would be no reason to call it meaning,

Fig. 6.4 The three elements of semiosis, according to Prodi

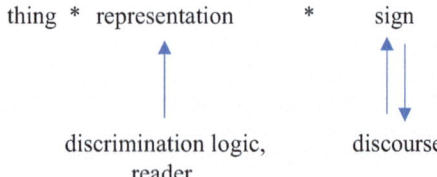

since it would be a thing like many others. Prodi's goal is to give a naturalistic account of meaning without giving up the *semiosic* and *semiotic* reality of meaning. If Prodi had limited himself to equate the meaning of semiosis—in its most complex forms, like verbal language—with either the referent of the sign or with the mnemonic trace in the brain, he would simply have argued that meaning, strictly speaking, does not exist. But this would not be a naturalistic explanation of semiosis: it would simply be the elimination of semiosis from the set of natural phenomena. Here we are here faced with the problem of understanding meaning *qua* meaning. When thinking about the notion of "meaning", Prodi almost certainly had in mind Frege's fundamental distinction between sense [*Sinn*] and meaning [*Bedeutung*]. Let us consider Frege's famous example: the two statements "the morning star is a body illuminated by the Sun" and "the evening star is a body illuminated by the Sun" have the same meaning, that is the same referent-thing, since they both are talking about the planet Venus. However, the two clearly have a different "sense" (Frege 1984: 162). Sense and reference are semiosic entities, and they should not be confused with the ideas that someone who uses a sign could have about its meaning: "the meaning and sense of a sign are to be distinguished from the associated idea. If what a sign means is an object perceivable by the senses, my idea of it is an internal image [...]. Such an idea is often inbued with feeling [...]. The same sense is not always connected, even in the same man, with the same idea. The idea is subjective: one man's idea is not that of another" (Frege 1984: 159–160). For this reason, meaning cannot be just the idea, or mental representation, of the user of a sign. The meaning of a sign, following again Frege, is "objective", that is it does not depend on the subjectivity of the user. If meaning was subjective, everyone would express his or her own meaning, and reciprocal comprehension would be impossible. This is why we need a codex, in order to make communication possible. But how is such a codex formed?

For Prodi, the third level of semiosis between sign and thing is where it is possible to find a *representation* that is "the internal analogue of the thing: it has all the external characters of translation in well-defined translation chains, that terminate in a natural state which is 'congruent' or complementary with the thing; this congruence is categorial, with the smallest number of categorial traits to permit the recognition of the thing" (Prodi 1982: 173). But a representation, following Frege, cannot be private; a representation is either intersubjective, or it is not a representation at all. For this reason the third level of semiosis is where a "codex" mediates between reader and thing-read. Let us take a closer look at a representation or meaning. Prodi is a coherent anti-substantialist (like Peirce and Wittgenstein before him), and therefore he does not take a representation to be a special kind of mental entity but rather a set of operations of translation chains that allow to identify it as different from all other representations: every representation is individuated, thanks to the set of operations that distinguish it from other possible representations (it is nothing but the relations it can partake to but also the relations it *cannot* partake to). This characteristic of representations accounts for their most important feature: they are not directly linked to the things they represent. Prodi indeed defines representations as "artificial analogues", that is to say "a stratum of the structure of interpretation that constitutes the [internal] translational analogue of a categorial reading" (Prodi 1982: 91). At this point natural semiosis can act onto the world in an "off-line" sort of way, since a representation permits the detachment of the reader from a direct link with the thing-referent:

> [w]ith the construction of artificial analogues it becomes possible to go from a purely con-sumptive process of knowledge-selection — where categorial recognition occasions a meta-bolic process that makes the survival of the reader possible — to a non-destructive categorial condition, that is to say one of storage and availability of "internal signs", or analogues or referent-signs, to be employed in further levels of reading by the reader that stored them, and whose semiotic interpretation will be conducted in the following stage. (Prodi 1982: 91)

Going back to Fig. 6.4, this means that the representation, at the most complex level of semiosis, has now become (relatively) autonomous from the thing it derives from. In the diagram presented in Fig. 6.2, the meaning of the sign coincides with the referent. Now this close link with the thing becomes unnecessary, since the "artifi-cial analogue" rests on the network composed by all the other representations:

> a signifying interaction [between reader and sign] is interpreted as expressing the fact that propositionality fabricates, with its mechanisms, an *ad hoc* class. It is as if, having been given a key, the logical machinery was able to use available and pre-made pieces in order to build a suitable lock. (Prodi 1982:91)

The translation chains triggered by the interaction with a sign do not need to rest upon a pre-existent mental content; the content of the sign is a kind of "self-categorization". A category that makes itself out of pre-existing categories. Consequently, proper lin-guistic meaning is a "translation network" (Prodi 1983b: 190).

Here the question of the codex emerges, which guarantees the intersubjectivity of knowledge and of communication. We have seen that, for Prodi, it is always nec-essary to avoid splits in the biological process. Biology means continuity. But, in this case, it is necessary to move from what happens to a single thing-organism to what happens in a community of things-organisms. The problem is that the codex— no matter how innate the faculty of language is (Bolhuis et al. 2014)—is learnt. So far, Prodi has only offered an account of either semiosic phenomena taking place within the single organism, or of innate semiosic phenomena, like the dance of the bees. The first feature of a codex is that of composing a "closed system":

> the jump [...] occurs when the internal analogues are so thoroughly interweaved (so, con-cretely, when an adequate neurological structure gets formed) that they can constitute a closed system, within which autonomous simulations are made possible that have as signs, or signifying states, the internal analogues themselves. We therefore move from a state of openness towards the outside — being dependent on the outside, even if through very long processes of mediation — to a state of self-closure, where the operations of translation can both begin and terminate on the inside. (Prodi 1982: 114)

The semiosic system thus gets sufficiently complex and intertwined to achieve inde-pendence from the conditions from which it derives. Signs can now refer to other signs, and a direct link with the external world of things is no longer necessary. Prodi does not explain in greater detail how it is possible to move from this kind of individual system to a codex valid for a community of systems. Besides, this is what

Chomsky and others argue when talking about "a recent and rapid evolutionary emergence of language" (Bolhuis et al., 4). Prodi, essentially, seems to hold a similar thesis, although reformulated in his own theoretical vocabulary. It is not a matter of establishing what comes first: the individual brain, capable of using a language, or the codex allowing the single human beings to communicate and comprehend each other. Prodi's solution is to propose what we would today call the co-evolution of mind and language (Deacon 1997; Aboitiz 2017). This is a solution that showcases the virtues of the model of the circle. Let us take the case of the linguistic "rule". A linguistic expression that does not presuppose a rule would not be intersubjective: but how—and why—could a rule without a linguistic expression have formed? Prodi shifts the problem, considering both of its poles the same time—the creation of a rule:

> is not an *a posteriori* with respect to the sign. It is not the case that there would first exist a rule that could *then* be applied. The rule is formed. […] The codex emerges contextually to signs (and not after the signs), and contextually to the cerebral-structural modifications that make signs, and their operations, possible. The codex is therefore essentially a product of the auto-compatibility between signs, and of their functional use in human phylogenesis. (Prodi 1982: 180)

Such "closed system" is a precondition for the formation of a codex which, in turn, is a precondition for the formation of a "closed system". Once the codex is formed, the way the human reader relates itself with the world changes radically. From the moment signs can be formed out of other signs, it becomes possible to think about something that does not exist. For Prodi, the linguistic sign always stands for a hypothesis. According to Fig. 6.4 (and reading from left to right), the thing comes first, but it is also possible to proceed from the other direction, from the sign to the thing. Language therefore establishes a "simulated use of the world" (Prodi 1987b: 153). Thus, scientific knowledge too—a systematic praxis of hypothesis-formulation and experimentation—can be explained as a form of natural semiosis. Indeed, the premises of scientific reasoning are already implicit in the diagram in Fig. 6.2: scientific knowledge "presupposes the whole of evolutionary history" (Prodi 1974: 29). So, for example, the bacterium's action of "searching" for the nourishment it needs already displays an implicit hypothetical character. This applies even more pertinently to human semiosis:

> every human sign is hypothetical, in the sense that it constitutes a mental fact derived from acts of connection between things and events. The sign does not exist, in nature, outside of the act of interpretation, which for complex readers means a postulated, hypothetical, partial, revisable, and testable reconstruction. For this reason, the sign "horse" is just as hypothetical as "unicorn". (Prodi 1977: 137)

This is a very interesting and original argument. Prodi's radical commitment to continuity does not at all entail a denial of the specificity of human cognition. That the linguistic sign "horse" is just as hypothetical as the sign "unicorn" means that the human animal *always* exhibits a hypothetical stance towards the world. To derive from the natural world does not mean distancing oneself from it. Prodi never relinquishes his role as a naturalist scientist and philosopher, even when he acknowledges how *Homo sapiens* has evolved far from its biological origins. Once again, the key is to take seriously the notion of biological continuity, something that does not at all exclude the possibility of abrupt evolutionary changes:

> evolution introduces qualitatively new systems. Propositional systems are wholly peculiar to man. A radical biological interpretation, like the one proposed here, does not at all correspond to now-fashionable forms of biologism,[3] according to which the facts of man can be uniformly and continuously reduced to previous facts. (Prodi 1982: 85)

This is the theorethical goal of Giorgio Prodi, to rebuild the steps from the bacterium to the unicorn, that is, from the very first forms of knowledge to the hypothetical ones. However, if the "unicorn" is a hypothesis, it is an attempt to grasp and comprehend the world, the world of things; therefore the evolutionary sequence is that from things to the bacterium, then to the unicorn, only to then go back to the things. Prodi's biosemiotic ontology—which is based on the model of the circle—describes a continuous return of all entities to their starting point, yet enriched by all the intermediate steps already crossed; a starting point that is represented by the things, or better yet, by things understood as a network of relations, as life itself (Barbieri 2006). Semiosis emerges out of selective interactions between agents, which categorize and "select" each other while ignoring all others: semiosis emerges from things, and it eventually returns to things (and therefore back to the starting point, in a circular motion). Reaching a conclusion that will appear paradoxical to those who believe that to be a materialist necessarily implies the abandonment of language and the mind as real and natural phenomena, this means that "if the representation 'stands for' the thing" and the sign "'stands for' the representation, and therefore the thing", then the sign is, for a reader, "equivalent to reality" (Prodi 1982: 175). Semiosis is as real as the world of things, because it is nothing but a transformation of worldly things. At the human level, this means that language is as natural as a flower is real for the bee suckling its nectar or the air for a seagull in flight.

If the foundational trait of Prodi's semiotic ontology is *continuity*—taking the shape of indefinitely extended and interweaved "translation networks"—then this is an unavoidable conclusion, since things are translated into signs and readers of signs. There is no substantial difference between things and signs: "thing, representation, and sign are one the mirror of each other, as we have seen regarding translation chains. But with regards to the propositional reader, thing and sign are (from certain points of view) equivalent, being both equally evaluated (as peers) by the reading system" (Prodi 1982: 176). A representation is not a separate entity from the mind of the reader: it is nothing but a "convergence" of a multiplicity of translation chains. Therefore, it will never be able to claim any autonomy—of either transcendental or cognitive variety—from the things that can be known through it. A representation is

---

[3] Prodi here is probably thinking about Wilson's (1978) sociobiology. His *La storia naturale della logica* was published only 4 years later: "to think that human behaviour can be deduced from animal behaviour, as believed by much of experimental psychology and sociobiology, is conceptually erroneous. The animal model is certainly useful, but inadequate. The human psyche was built in a very peculiar manner, precisely thanks to its intimate mingling and merging with logico-discursive functions" (Prodi 1987: 61).

nothing but another—internal, mental—form of the things themselves. And since the reader himself is nothing but a federation of signs, and the things are intrinsically signs, at the end of this chain of equivalences the 'reality'—for the human reader as the most complex and articulate of readers—cannot be distinguished from the signs we use to think and to describe it. For this reason, since at the end of the semiosic process we also find its origin, "things make themselves be read and spoken about" (Prodi 1983b: 192). This last remark allows us to return to Fig. 6.4, in order to offer a more precise interpretation of it. We have already noted how Prodi disliked the semiotic triangle, and we are now in a better position to understand why. The diagram in Fig. 6.4 is not, properly speaking, another version of the semiotic triangle. While the latter presupposes three *distinct* entities that come into contact, in Prodi's biological hypothesis, the three entities do not subsist as separate entities, they are rather the transformation of one into another, or better yet they are different forms—in evolutionary and cognitive terms—of the *same* process, life as semiosis.

Once again, the circle comes to a close: from things to signs to representations and back again. The semiosic problem that we have presented at the beginning of this chapter—how to justify the link between sign and referent—is now completely gone. If there is a problem left to be addresses, it is only that of the *apparent* separation between thing and reader, between sign and referent, and between representation and world. The union of the two entities is not a problem, since the sign—as a biological entity—is born as a unit, as an assimilation of reader and object. The separation between sign and referent then is never actual: it only appears to exist because the translation chains that link the two are very extended, to the point that the primordial spatial contiguity between the two becomes undetectable. But if the core feature of semiosis is continuity—the model of the circle once again—then there will never be a radical separation between sign and referent, and this flattens the semiotic triangle to a continuous line of transformations of entities into more complex entities.

> [t]he double face of the sign is given the moment that, in nature, a sufficiently complex codex appears. But if we could deconstruct a sign into its various steps, we would discover [...] that every step is formed by a reader-thing that reads another thing, which thus becomes a sign. And this reader-thing, in turn is read by something else, and so on. All the molecular processes that we are starting today to understand sufficiently well (from the synthesis of proteins to the use of energy, from the duplication of DNA to the transmission of nerve impulses) are of this kind. The cerebral function of man is stupendously complex [...] but it is hard to imagine that it would function in a radically different way from that of a continuous molecular acknowledgment of meaningfulness. (Prodi 1987b: 147)

Let us conclude this long chapter, probably the most important one in order to fully appreciate Prodi's project, by returning again to Prodi's relationship with biology. Prodi, as we have seen time and again, semiotizes biology, just as be biologizes semiotics. For Prodi, life is intrinsically semiosis. That is to say, he merges together two domains that—at the time he was writing and even more today—are often kept well separate: biology and nature on the one hand and semiosis and culture on the other. Prodi did not believe in this separation (which corresponds to that between natural and human sciences) and always strived to identify their point of contact. Here "point of contact" means a point that would be located *beyond* the distinction between nature and culture, without being either just nature or just culture. This is where the identification of life with semiosis takes place. It is not surprising that this was also Darwin's choice, who proposed to "read" the natural world as a book, albeit an incomplete one—to read it as a semiosis:

> I look at the geological record as a history of the world imperfectly kept, and written in a changing dialect; of this history we possess the last volume alone, relating only to two or three countries. Of this volume, only here and there a short chapter has been preserved; and of each page, only here and there a few lines. Each word of the slowly-changing language, more or less different in the successive chapters, may represent the forms of life, which are entombed in our consecutive formations, and which falsely appear to us to have been abruptly introduced. (Darwin 2006: 289)

Chapter 7
The Origin of Language

> The separation between the biological and what is called the
> "spiritual" [...] can be interpreted in two ways. The spiritual
> could be thought of as too complex to be explained with the
> vocabulary of the biological, and the biological too rough to be
> capable to explain that which is spiritual. [...] These are,
> clearly, two formulations of the same proposition. One
> emphasizes the beauty and the perfection of the spiritual — its
> non-naturality. The other emphasizes the mechanical character
> of biology. [...] I have preferred to take a different path, one
> already looking for some kind of intelligence (not of human or
> anthropomorphic fashion) in the biological, and considering
> every complication — including logic and rational discourse —
> as a complication of this intelligence. I called this stance
> "natural rationalism", identifying it with the elementary
> semiotics that lies at the foundation of every biological
> organization.
>
> (Prodi 1989: 94)

Abstract How does the transition between the proto-semiosic material relation, established between two molecules, and a historical-natural language happen, and what changes does it bring? How much in common do cellular semiosis and human semiosis have? It is necessary to understand how Prodi explains the evolution of semiosis, from its most simple forms to the more complex ones. The problem is how to maintain the continuity of the biological process without underplaying the radical discontinuities it constantly engenders.

Keywords Evolution · Discontinuity · Transformation · Origin of language

Let us return to the circle, the basic model—in our interpretation—for Prodi's semiotic ontology. To consider this as the foundational model of semiosis means to assume the principle of continuity as the distinctive trait of all relations established between different forms of semiosis (and, therefore, forms of life) which, in the circle, succeed one another through an unbroken circular process. We immediately

© Springer Nature Switzerland AG 2018   69
F. Cimatti, *A Biosemiotic Ontology*, Biosemiotics 18,
https://doi.org/10.1007/978-3-319-97903-8_7

face a problem: human semiosis is continuous with the forms of semiosis that preceded it—and particularly with various forms of phytosemiosis (Krampen 1981) and zoosemiosis (Sebeok 1968; Cimatti 1998; Maran et al. 2011)—but, at the same time, it also represents a radically different system. That is to say, it represents a discontinuity with respect to non-human forms of communication. One has to note that the use of terms such as "continuity" and "discontinuity" simply aims at emphasizing, on the one hand, the general biosemiotic characteristics that proto-semiotic and human language have in common; on the other, there are also some characteristics of human language which are not in common with "simpler" forms of semiosis. How, then, can we reconcile continuity and discontinuity (Eldrege and Gould 1972; Sheldon 2001; Hunt et al. 2015)? In reality, this couple of terms appears to be mutually alternative only when superficially examined, since *every* form of life on this planet is, at one and the same time, in a relation of both continuity and discontinuity with those who preceded it: continuity, because otherwise it could not exist, since no form of life emerges out of nothing, but it is always a transformation of other forms of life; discontinuity, because every form of life is adapted to a *particular* environment, that is to say that it has specialized itself (quite literally: it has become a species) with respect to that *environment*. This means that any form of life has something in common with the forms of life which preceded it in time, but that it has also some specific characters that it does not share with the other forms of life. The problem emerges from the fact that continuity seems to imply gradualism. "Continuity" means that there are no gaps between various forms of life, since they seamlessly undergo transformations in the evolutionary process. For example, at time *t* the form of life *B* is linked—through a continuous series of transformations—to form of life *A*, now extinct and which lived at a previous time. Gradualism, on the other hand, means that these transformations need to be extremely small such that, for example, there will be only imperceptible differences between form of life *A* at time *t* and form of life *A-1* at time *t-1*.

But evolutionary continuity does not at all imply gradualism. From a rigorously biological perspective, the alternative between continuity and discontinuity is unsatisfactory:

> because the progressive complication of functions can translate — when reaching a certain threshold and meeting some precise requisites — into a critical restructuring (so to speak), i.e. a functional-structural novelty: human linguistic abilities can be explained by some growing interpretive capacities which, once reached a given threshold, mutually interact and give rise to a wholly new situation which is, however, still explainable as a gradual evolution of its single components. (Prodi 1983b: 180)

Prodi is primarily interested in human language. It is important to remember that Prodi was writing in a period during which Chomsky's generative-transformational linguistics was profoundly influential. Chomsky's point of view is interesting and often misunderstood. Obviously, he admits the existence of communication systems in non-human animals. But at the same time, he observes how their characteristics are very different from those of human languages. This does not imply that zoosemiotic codes are less meaningful than human language, nor does it entail a "devaluation" of the world of non-human animals. Chomsky, like Prodi, is actually as attentive to similarities as he is to differences. In fact, it is not clear why Chomsky's stance has been so often—and so fiercely—criticized. For example, this is what he has to say, in his *Managua Lectures*, regarding the specifically human capacity of "discrete infinity":

> [t]o put it simply, each sentence has a fixed number of words: one, two, three, forty-seven, ninety-three, etc. And there is no limit in principle to how many words the sentence may contain. Other systems known in the animal world are quite different. Thus the system of ape calls is finite, there are a fixed number, say, forty. The so-called bee language, on the other hand, is infinite, but is not discrete. A bee signals the distance of a flower from the hive by some form of motion; the greater the distance, the more the motion. Between any two signals there is in principle another, signaling the distance in between the first two, and this continues down to the ability to discriminate. One might argue that this system is even "richer" than human language, because it contains "more signals" in a certain mathematically well-defined sense. But this is meaningless. It is simply a different system, with an entirely different basis. To call it a "language" is simply to use a misleading metaphor. (Chomsky 1988: 169)

Chomsky's thesis is rather simple and uncontroversial: compared to human language, the bees' communication system is "simply a different system, with an entirely different basis". Chomsky proposes a technical definition of human language, a system capable of "discrete infinity"; a set of rules of communication that does not have this capacity is, by definition, not a language. What is so scandalous about this? The most interesting point is that, for Chomsky, the ability to communicate is not the distinctive feature of human language. Language would more properly be conceived as a cognitive—and in particular, arithmetical (Chomsky 1988:169)—tool, rather than one meant for communication: humans use language for thought more often than they do for communication. The fact that a cat does not possess the ability of "discrete infinity" is no more surprising than the fact that a human is not endowed with wings. As Darwin reminds us, life means *diversity*. Just as not having wings does not mean that the human body is inferior to that of a seagull, so the lack of "discrete infinity" does not entail that the cat would be less intelligent than a human being. Just like there is more than one way to move through space (and one method is not necessarily better than all others), there is more than one form of intelligence: different animals in different environments mean different minds. From this point of view, Chomsky is thoroughly Darwinian: "there is a long history of study of origin of language, asking how it arose from calls of apes and so forth. That investigation in my view is a completely waste of time, because language is based on a entirely different principle than any animal communication system" (Chomsky 1988: 183). Evolution does not imply gradualism, since it does not exclude biological discontinuity (Eldredge, Gould 1972). At the same time, such a position seems to adhere strictly to Jakob von Uexküll's basic tenet that every form of life can only experience its own *Umwelt*. This means that an "abstract" comparison between the cognitive capacities of different forms of life—that is, animal species living in different *Umwelten*—is meaningless. When Chomsky says that the language of bees is different from human language, he is saying simply that *Apis mellifera* and *Homo sapiens* live in different *Umwelten*, characterized by different semiotic properties. This is normal science.

Prodi's position is not at all far from Chomsky's. To build an evolutionary link between language and the biological world does not mean to deny that the former possesses some unique features, which cannot be found in other forms of semiosis. But let us begin with the similarities, in particular at the cerebral level. Prodi writes:

> there are no […] specific "atomic" mechanisms (i.e. elementary and constitutive), that is nerve structures or functions that would be exclusive to man. The system of conduction through depolarization is the same (it is a constant, a 'biological universal', at least from a certain level of complexity upwards), the neurotransmitters are roughly the same such that there is no uniquely 'human' neurotransmitter, and there is no 'human' nerve tissue. Even the types of synapses are the same as those that can be found in animals. (Prodi 1987b: 52)

Thirty years have passed since Prodi wrote these lines, and the question of the similarity of the human brain with that of an anthropomorphic monkey remains highly controversial (Rilling 2006; Marks 2015; Vermunt et al. 2016). A general point still stands. The clear similarities between human beings and monkeys—and more generally those between *Homo sapiens* and the rest of the animal world—make it possible to claim that human behaviour, and human language in particular, has no close analogues in the natural world. How, then, should we explain this characteristic?

According to Prodi, human language exhibits at least three species-specific characters (i.e. features capable of singling out the specificity of the human): (1) "[language] is founded on a logical competence, part and parcel of the linguistic function. For this reason, human language can be considered essentially human" (Prodi 1983b: 173). (2) In our species, language coincides with our being animals "humanization […] means 'origin of linguistic competence in nature'" (Prodi 1987b: 69). Finally, (3) the evolutionary environment of the human animal is constituted by language itself; "it is impossible to overstate, in order to explain the genetics of communication (i.e. the fact that the human has, in its genetic code, the instruments for communication) the importance that communication itself has played in modelling its structure, through selection, and how this has perfected communication itself: a virtuous circle, acting through a dense network existing above single individuals and generations alike" (Prodi 1977: 143).

The first point establishes that "propositional logic"—founded upon the "simplest" norms of natural logic (both "material" and "categorial" logic) yet still distinct from these because of its generality and its combinatorial capabilities—in fact *coincides* with human language: "indeed language is nothing but a natural situation, acquired *after* — yet together with — propositional logic. Moreover, it is not a function external to propositional logic, but it mostly coincides with it" (Prodi 1982: 166). However, for Prodi this does not mean that, in the human animal, thought and language are one and the same: "language is at the same time richer and less precise, and therefore it contains more than logical operations and their relation to the world" (Prodi 1982: 167). What matters is that the emergence of the ability to use a complex language modifies the very cognitive abilities that have permitted its development. Once again, referring to the model of the circle helps us understand Prodi's stance. The more or less "sudden" appearance (regarding the timeframe of biology, see Bolhuis et al. 2011) of the capacity to use a complex set of rules has a retroactive effect on the mental characteristics that facilitated the development of that very set of rules. That which follows can modify what came before. For example, consider how the thought of a human animal changes after it becomes able to think through the signs of its language (Cimatti 2000c). As we have seen in the previous chapter, every sign is a hypothesis. This means that to think *through signs* entails to be able to see the world not just as it is now but as it could be. Going back to Prodi's example, every "horse" is *also* a "unicorn": a non-existent, fantastic animal that nonetheless *could* exist (Lavers: 2014). The human scientific and experimental attitude begins here:

> things do not reveal themselves integrally, they do not show up in their fullness, but give incomplete signs of a partial and elusive presence. These signs must be searched for and confirmed, as well as continuously kept together, in order to oppose their tendency to dissolve. (Prodi 1974: 24)

An animal capable of having these thoughts is, at *the same time*, an animal just like any other primate and completely different from any other animal. If any sign is also a "hypothesis", then the world becomes something problematic, not to be taken for granted. Jakob von Uexküll's closed functional schema (Fig. 4.2) is thus broken. Now a non-metabolic use of the environment becomes possible, as Prodi puts it, that is to say a non-functional, "uninterested" relation with the surrounding environment. As we will see in the next chapters, it is in this suspension of the metabolic function that we should look for the foundation of the human animal's aesthetic and ethical capacities. In particular, Prodi considers language to be a specific form of *thought-action*. To speak a language primarily means to think in a different way as compared to another animal—the latter might be able to communicate by means of such a language; however, it is unable to *think in* such a language. The larger (relative) dimensions of the human brain are, for Prodi, made necessary by language: "in the human brain a stupendously large number of synapses are confronted and selected by logical-linguistic facts, that likely occupy all the extra space that the human brain has available, as compared to its predecessors" (Prodi 1987a: 82). Such a brain, formed on and together with logical-linguistic competence, acts onto the world in a peculiar manner. Every action is redoubled: on the one hand, it is directed towards its goal, while on the other it implicitly contains a hypothetical network of connections that action could have with other possible actions. Human actions are, typically, "linguistic actions", that is to say actions that would have been impossible without logical-linguistic thought:

> the individual enters into the linguistic fabric by means of logical reactions. [...] It is clear that his particular logic cannot react with material presences (like when the eye opens to the light) but with logical presences, i.e. logical structures in the world — and these cannot but be linguistic. (Prodi 1987a: 84)

If every word is a hypothesis, then every thought/word makes an action possible that is not simply dependent on what actually exists in the world; if the unicorn does not exist, maybe it is possible to invent it: "in the linguistic fabric, aside from language, there are a large number of so to speak 'linguistic' things, that are as concrete as the things of nature, and just as binding" (Prodi 1987a: 85). The human world is the world of "linguistic things", a world where the unicorn is as real as the horse.

Language then becomes the primary engine of human evolution. The human species is not, from this point of view, an animal species like any other, where semiosic competence runs parallel to other abilities, like perceptual or logical competence (categorial logic). The human animal is language itself:

> for man, this specificity is so profound to constitute a qualitative jump: the discursive ability, that is the use and interpretation of reality through the mediation of words, along with the necessary implied logic. This is the specific problem of human biology. (Prodi 1989: 92)

A non-human animal is adapted to its environment, and, in addition to this, it is able to communicate; in the case of the human animal, on the other hand, this distinction between two levels—the natural level and the semiosic one—or between environment and communication no longer applies. *The environment of the human animal is language* itself: the human animal is adapted to language; it is made *by* and *for* language. Even if according to Prodi biology means semiosis, *human* biology is a "second-degree" semiosis:

> [m]an, like any species, has its own specific characteristics. The peculiarity of man [...] is speech. Animals are adapted to things or to defined presences, and this adaptation is of course a form of communication with these things: life as a whole is a form of communication. The ant eater is in communication-adaptation with the ants, the herbivores with the plants, etc. But man is adapted to communication itself. His is a second-degree communication: a symbolic and rational competence. So, man is not evolutionarily selected by things (by nature, if meant in rough material sense), and we must not conceive him as adapted to things, but rather to relations. (Prodi 1987b: 70)

The intrinsic logic of semiosis and categorization—the repetition of cycles of interaction with the environment, as exemplified by Fig. 6.4—unintentionally produces ever more complex systems, as for example an animal gifted with semiosic and cognitive abilities similar to those of contemporary primates. At this evolutionary stage—probably following the dramatic quantitative expansion of our brain as compared to other animals (Knight et al. 2000; Lanyon 2006) that allowed for the development of the capacity of enormous combinatorial possibilities—the evolutionary path of our species has turned towards language. If the anteater, to return to Prodi's own example, has adapted to its environment by developing an elongated snout and a thin, extensible tongue in order to catch his food, the human animal has exploited this sudden "excess" of neuronal resources—the fortuitous and lucky result of a favourable mutation which, for example, could have dilated the times of ontogenesis, a characteristic linked to neoteny, "a slowing down of ontogeny and retention of previously juvenile stages as adult forms of descendants" (Gould 2002: 369)—in order to better adapt itself to language, i.e. to use language with ever-increasing efficiency and versatility:

> how can man be adapted to an artificial object (language) which he himself has built and that, therefore, cannot precede his existence? The answer is, as usual, to imagine simple situations becoming more complex, or "self-complicating" themselves. It is not possible to think of a language emerging all of a sudden, fully formed. It is rather necessary to imagine that very elementary forms of communication or language (at the level of zoosemiotics [...]) have constructed a selection paradigm for *genetic* communication capacities of a slightly higher level, and that these, in turn, have created more evolved forms of communication — more exigent, more selective — and again that these have selected more *sophisti-*

cated genetic skills. [...] By following this model, language has selected itself. (Prodi 1987b: 70)

Let us consider this final claim with the utmost seriousness: "language has selected itself". Prodi is saying that the appearance of language has changed the evolutionary path of the human species. All of a sudden, it no longer was a matter of direct adaptation to the natural world, the climate, predators, social life, and so on. These are common constraints for all forms of life. Instead, at a certain point during its evolution, *Homo sapiens* started adapting to the primary tool for the "construction" of its environment: language. For example, language made possible a very rapid technological development, far faster than that allowed by learning through imitation (Galef: 1988). A technology "based" on language quickly becomes widely diffused and long-lasting. It does not need to begin anew with each new organism (this is the so-called ratchet effect; see Tennie et al. 2009). It is fully plausible to imagine a genome, making possible the efficient use of language, would be "preferred" by natural selection, considering the great adaptive advantages it brings to the species. This means to adapt to one's own behaviour. More precisely, it implies a retroaction on the very same genetic preconditions that allowed the development of that adaptive kind of behaviour in the first place. This means that behaviour, the phenotype, somehow "drags along" the evolution of the genotype. This general pattern of action can be found in all life forms. Indeed, every organism modifies its environment—in a more or less profound way—and this modified environment, in turn, applies an evolutionary pressure on the species as a whole. This is the so-called mechanism of "niche construction":

> (1) organisms modify environmental states in nonrandom ways, thereby imposing a systematic bias on the selection they generate, and allowing organisms to exert some influence over their own evolution; (2) ecological inheritance strongly affects evolutionary dynamics, and contributes to parent-offspring similarity; (3) acquired characters and byproducts become evolutionarily significant by affecting selective environments in systematic ways, and (4) the complementarity of organisms and their environments (traditionally described as 'adaptation') can be achieved through evolution by niche construction. (Laland et al. 2016: 192)

This "niche construction" applies to every form of life, but it is particularly effective for the human species. The beaver's environment is the dam it has built. But the human's environment is not an "external" artefact; it is rather an internal tool—language—that allows the species to formulate hypotheses about any other artefact. The human species' environment is not the dam in the river, like it is for the beaver, but, so to speak, a dam built in its own brain. Ultimately, to be adapted to language means to be adapted to no environment in particular, precisely because through the use of language—and particularly by means of hypotheses—it is possible to construct any environment one desires. Therefore, human animals suddenly adapted to their own behaviour (the phenotype becoming a propulsive force with respect to the genotype): indeed, to the most important behaviour, granting *Homo sapiens* an extraordinary evolutionary fitness. However, we need to keep in mind that when Prodi speaks of human language, he is not thinking of communication or semiosis in general—phenomena that, as we have seen, do not begin with *Homo sapiens*.

Instead, Prodi thinks of language qua "discourse" or "logical network" and mostly qua "syntax", and in this case Prodi cannot but be Chomskyian. Indeed, syntax means the capacity to combine together new and unprecedented thoughts: the syntactical-combinatory capacity gives the human animal an infinite repertoire of hitherto unthought thoughts. In the human world, there is no hypothesis without syntax and, therefore, without language:

> [d]iscourse [...] is the elaboration of propositional logic for actions upon the world, i.e. performed to interpret and modify the world. Categorial knowledge is not given outside of interlinked translation chains. But in the case of propositional knowledge there is a more urgent necessity to move onto things and to adapt to them. The hypothetical strategy can connect — on the base of antecedents — natural facts that are very far one from the other, making them collide. [...] Therefore, the transformation of the world operated through linguistic mediation is intrinsic to knowledge: the latter needs to constantly prove itself, as derived from the real and adapted from the real. From this point of view, propositional logic and discourse too are great empirical systems, they are *a posteriori* of phylogenesis and themselves forms of phylogenesis, that is to say, of evolution and change. (Prodi 1982: 195)

"Discourse" is an a posteriori, it is a phenotypic manifestation of a genotypic character: it is an effect of the a priori and innate capacity of using language in a productive and hypothetical manner. However, this a posteriori makes possible the construction of a world (technology, culture, traditions, etc.) that in turn exercises an influence on the biological a priori (the is, on the genotypic character). For example, every genetic modification that made learning a language an easier task (Deacon 1997) would have been immediately "preferred" by evolution, since it would have made language acquisition and use—the principal "engines" of human development—more efficient. It therefore becomes possible to hypothesize a co-evolution between language and the human species (Cavalli-Sforza et al. 1992; Briscoe 2002; Pakendorf 2014). A co-evolutionary relationship between the development of human semiosis and propositional language (a complication of pre-existent forms of animal semiosis) and cultural development *qua* further evolution of the genetic basis that made possible the "logical network" of the syntax of human languages. It can therefore be postulated that:

> intersubjective communication was the new selective cultural condition. In other words, the first 'external' selective factor was constituted by the first rudimental forms of communication, and the most efficient readers were selected through the advantages given by communication to those who made use of it. Therefore, communication was both selected and constructed by communication itself. (Prodi 1982: 166)

The retroactive effect of language onto its own a priori conditions of possibility, and the co-evolution of language and *Homo sapiens*, allows Prodi to both uphold his belief in continuity—since human propositional language could not have existed without natural biosemiosis—without renouncing the 'unicity' of the human species with respect to the rest of the animal world (Cimatti 2015). Thus, Prodi can highlight a certain discontinuity without however endorsing the dualism implicit in Cartesian semiotics:

> [w]hen speaking about biology, it is common to refer to entities and phenomena such as digestion, hormones, pulse, molecular biology, and so on. Language and thought are considered to belong to a different chapter. [...] On the contrary, I believe that language and thought

belong to human biology, and indeed that they are its most distinctive trait. The fact that they are complex and currently obscure (in their belonging to biology and therefore being explicable in terms of molecular structures and suchlike) does not allow us to set them apart from biology and to relegate them to a separate chapter — one pertaining to the human sciences. Thus, man is continuous with respect to the other species, but not reducible to them. It is made in a radically different and well-defined way, although this difference is produced by a rearrangement of already-existent facts. (Prodi 1989: 92–93)

Prodi wants to avoid underestimating the novelty of human language as compared to zoosemiotic systems—that is, looking at it merely as a "complication" of very similar pre-existent phenomena. But, at the same time, he is wary of overestimating the unicity of human language as compared to the characteristics of natural biosemiosis. The first pitfall is that of seeing language merely as a means of communication and the second that of seeing it just as syntax. In the first case, the similarities with the infinite number of other systems of animal communication are evident. In the second, it is clear that something as complex as the syntax of human languages cannot be learnt but must be innate. In both cases, we lose sight of the most distinctive feature of human language, the intertwining of biological/innate aspects and of cultural/historical ones. Language is neither fully historical—because it wasn't created by humans—nor simply natural, since it is also true that without the active and constructive role played by the linguistic activities of human animals, there would be no the actual difference of human languages. The human animal is both cause and effect of language, and this means that language is the defining characteristic of our species: "human capacities for knowledge, that manifest themselves well before scientific knowledge, are based on language, i.e. on that which is specific (species specific) to man" (Prodi 1987b: 70). Here Prodi is adopting Darwin's still-timely stance, for when the latter—in his *On the Origin of the Species*—examines human language, he describes it as somehow similar to animal semiosis without underplaying those features that radically set it apart from non-human systems of communication. In this analysis, the starting point is the acknowledgement that "articulate language is, however, peculiar to man" (Darwin 2009: 54). What is at stake here is not simply the communicative/expressive power of language—which is also present in other animals—but the ability (a syntactic ability, as we would put it today) to form and express articulate and complex thoughts:

> our cries of pain, fear, surprise, anger, together with their appropriate actions, and the murmur of a mother to her beloved child, are more expressive than any words. It is not the mere power of articulation that distinguishes man from other animals, for as every one knows, parrots can talk; but it is his large power of connecting definite sounds with definite ideas; and this obviously depends on the development of the mental faculties. (Darwin 2009: 54)

The development of mental faculties proceeds in lockstep with the ability to form complex combinations of words. Mind and syntax are two sides of the same natural coin. Darwin's definition of human language—neither an "art" nor an "instinct"—is even more interesting:

> [l]anguage is an art, like brewing or baking; but writing would have been a much more appropriate simile. It certainly is not a true instinct, as every language has to be learnt. It differs, however, widely from all ordinary arts, for man has an instinctive tendency to speak,

as we see in the babble of our young children; whilst no child has an instinctive tendency to brew, bake, or write. Moreover, no philologist now supposes that any language has been deliberately invented; each has been slowly and unconsciously developed by many steps. (Darwin 2009. 55)

There is a natural (innate) tendency to speak, but languages are learnt. These, however, are not invented, like someone has invented the art of baking: there is therefore an "instinctive tendency to acquire an art" (Darwin 2009: 56). It is difficult, and perhaps pointless, to try and trace a precise boundary, into the human capacity to master a language, between what is innate and what is acquired, between what is natural and what is cultural, and between what is universal and what is historical. Human language, for Darwin, is this mixture of nature and history. Thus, the continuity of the evolutionary process is never questioned, but neither is the specificity of language underestimated. Indeed, Darwin explicitly hypothesizes a relation of co-evolution between the human brain and language, a thesis that, as we have seen, Prodi enthusiastically endorses:

The relation between the continued use of language and the development of the brain has no doubt been far more important. The mental powers in some early progenitor of man must have been more highly developed than in any existing ape, before even the most imperfect form of speech could have come into use; but we may confidently believe that the continued use and advancement of this power would have reacted on the mind by enabling and encouraging it to carry on long trains of thought. A long and complex train of thought can no more be carried on without the aid of words, whether spoken or silent, than a long calculation without the use of figures or algebra. (Darwin 2009: 57)

Just like the anteater is adapted to its environment, and in particular to the animals that are its prey, the human animal is adapted to language, of which he is both cause and effect. The anteater is adapted to something external to itself, while the human animal is adapted to something that is neither fully external nor internal and neither completely innate nor learnt and historical. Once again, we find ourselves to walk along the circle's circumference, where it becomes impossible to distinguish what came before and what after. Thus, the "improper" problem of the origin of language—improper to the extent that we seek an explanation for the ex nihilo (or almost ex nihilo, as gradualists hold) emergence of language—is fully dissolved once it is comprehended that there has never been, in our evolutionary history, a language-less human who suddenly decided to invent a language. That hypothetical human being, similar to us in all respects except his or her lack of language, has never existed: the human animal is what it is because it literally constructed itself around language:

[u]ntil now, we have simply acknowledged a peculiar way of being of the human: it cannot subsist in itself without this communion with others by means of culture as a fact of communication. The problem is to understand how we have come to this situation. If the frog is adapted to the fly, man is adapted to language. But the fly was not produced by the frog: if the frog can subsist by eating flies, that is because during natural history it has outsmarted it, it has used it, it has adapted to it — shaping itself onto the fly (and the grass, and the water, and so on). Language, on the other hand, was produced by man, and not by nature. [...] Therefore, it looks as if man has formed itself, through his natural capacity for language use, onto something he himself built. It is a snake eating its own tail. [...] It is there-

fore necessary to question this fundamental incongruence: man has adapted himself to language, but he also constructed his own language. [...] But if this mechanism works, that is if the linguistic mediation through which all men come to recognize themselves as men takes place, it is only because man's brain is suitably structured. The brain somehow fits the language, and the two mechanisms are a good match to each other. [...] How could this natural predicament occur? Likely thanks to reciprocal relations that were established over time. Let us assume then that not only man has constructed languages, but also that languages have constructed man — his specific logical competence and his fluency in discursive communication. (Prodi 1987b: 47–48)

It is here, in this complex tapestry, that the specific character of our language—setting it apart from other animal languages—should be sought. The specificity of human language is given by its coincidence with the animal who uses it, being the result of a process of co-evolution where the terms of the relation—language and human animal—have constructed each other. This solution clears the stage of all the distinctions that have been traditionally invoked to grasp the peculiarities of human language: first of all, that between natural and cultural elements. Language has both characters but does not fully coincide with either. In fact, if "natural" is interpreted as synonymous with "necessary", and "cultural" with "arbitrary" (since these are the automatisms suggested by the Cartesian framework), language is neither. It is not "natural" since its manifestation presupposes—for every new-born individual in our species—someone who can already use it, i.e. a historical community of speakers. Language is *learnt*, even though this learning process is facilitated by our genetic make-up. In this sense language is not an organ, like the liver, without which life would be impossible—yet without an innate predisposition to language, no particular language (like English or Estonian) could be spoken.

[l]anguage [...] based on logical-hypothetical functions, is specifically human. It has affirmed itself thanks to the very notable operative advantages it has given to the reader-man, i.e. to the new emergent species. But if things selectively constitute their readers, letting themselves be read, it is yet to be established against what natural obstruction (what kind of thing) linguistic competence — and the logical -propositional functions that make it possible — has constituted itself. It is certainly not natural reality, since this determines the categorial logic that serves as condition of possibility of logical -hypothetical functions. It is therefore necessary to postulate that the real obstruction against which linguistic competence has evolved was communication itself, that is to say, language. Postulating elementary forms of exchange, at the zoosemiotic level, we can hypothesize that they have selected ever more efficient forms of exchange, and that through this process language has selected itself. (Prodi 1983b: 197–198)[1]

---

[1] An important antecedent of this stance can be identified in the so-called Baldwin effect: ontogenetic behavioural changes contribute to the modification of the evolutionary environment of a species which, in turn, applies an evolutionary pressure onto the species' genome. A kind of Lamarckian effect takes place, although without necessarily entailing the inheritance of learnt behaviours. So, the species "cooperates" to the indirect modification of its own genome: "*the adaptations made in ontogenetic development which "set" the direction of evolution are novelties of function in whole or part* (although they utilize congenital variations of structure). And it is only by the exercise of these novel functions that the creatures are kept alive to propagate and thus produce further variations of structure which may in time make the whole function, with its adequate structure, congenital" (Baldwin 1896: 449).

This is a crucial point, since the development of a completely natural organ like, say, a kidney does not require the direct cooperation of other kidneys. Kidneys will effortlessly develop in the body of a child growing up isolated from other human beings. But that same child, growing up in an environment where language is not already present, will never spontaneously begin to speak. At the same time, language cannot be considered to be an arbitrary social institution, since it is bound by strict genetic constraints: indeed, our language cannot be taught to a non-human animal (aside from a minuscule and irrelevant number of words), even to those who are, by many criteria, very intelligent creatures (Lestel 1995). Similarly, a human animal who has grown beyond a certain critical age threshold will be unable to pick up language (Lenneberg 1967). There is continuity: language is wholly natural, since it coincides with the human environment. But there is also discontinuity, since *Homo sapiens* is one with language: "the [human] brain somehow coincides with language" (Prodi 1987b: 47). Prodi's position is at once paradoxical and very original. It is common to find researchers who link human language with that of other animals, but in so doing they underplay the specificity of historical-natural languages. It is just as common (although nowadays somewhat unfashionable) to find strong defences of human specificity, which amount to a claim of radical uniqueness. Prodi avoids to take either of these paths and remains a faithful Darwinian without forsaking Chomsky's ideas—believing in continuity without forgetting discontinuity.

> [t]he mediating tool that is language, employed by man to express all that is meaningful and representable, is profoundly different from behaviours dictated by plain and simple genetic rules [...] like for example those followed by the bees in order to establish their complex social hierarchy. While language is something existing "out there" — and as such it is a natural fact that has the same material concreteness as the flowers and the beehive — it is also a product of *other men* who have distilled a communicative function into a refined tool for communication — to this extent they are manifesting something in their genetic make-up, the impulse to communicate. Therefore, language is a culturally-manufactured natural thing, fabricated thanks to a natural (genetically impressed) competence. The combination of these profoundly different elements (like we still believe nature and culture to be) is very complex, and constitutes the defining characteristic of [...] the [human] species. (Prodi 1987b: 45)

Chapter 8
Attention and Consciousness

> *Dualism is not overcome by synthesis, but by the acknowledgement that, at the root of things, there is no dualism at all.*

> (Prodi 1987b: 119)

Abstract The human is an animal that refers to itself as an "I". According to Descartes, the subject is an axiom, and everything else follows from this primordial certainty. This is a dualism: to postulate an I as separate from the natural world. Prodi rejects this dualism. The challenge of Prodi is to find a naturalistically way to explain how human subjectivity can emerge from the world of things; that is, from biosemiotic complementarity to the I. For Prodi, following Vygotsky's hypothesis, the "I" *qua* self-conscious psychological entity, is inseparable from the pronoun "I", i.e. the discursive capacity to refer to oneself. Human consciousness is therefore the capacity to pay attention to oneself by means of language.

Keywords Anti-dualism · Consciousness · Self-consciousness · Attention · Vygotsky

The Cartesian model begins with consciousness, with mind. The I is an axiom. Therefore, Descartes' approach is dualist and radically anti-naturalistic. Conversely, as we have already seen in Chap. 5, Prodi's model is radically anti-dualistic: at the beginning, there is the relation between things and the world, natural semiosis. And yet, the experience of oneself as something interior and intimate seems wholly natural. It is therefore necessary to explain—like it was explained in Chap. 7 vis a vis language—how an animal capable of thinking of itself as an I could have emerged from the natural world. The topic of this chapter, then, is the natural history of human consciousness. This theme has recently received quite a lot of scholarly attention (Sherwood et al. 2008, Humphrey 1992; Tattersall 2016), but once again Prodi's solution is both profoundly original—especially so considering that it was formulated over 30 years ago—and coherent with his radical biologism. Let us

© Springer Nature Switzerland AG 2018

F. Cimatti, *A Biosemiotic Ontology*, Biosemiotics 18,

https://doi.org/10.1007/978-3-319-97903-8_8

begin with relations: these precede the consciousness of the I, as well as the very distinction between subject and object. Prodi's axiom is that:

> [a] properly interior knowledge (intended as an unveiling of something noumenal relative to thought) does not exist. This assertion needs to be clarified: we do have a kind of knowledge of ourselves as epistemic structures. But this is historical and progressive, correlated to our development and to our surroundings, and based upon — as well as sensitive to — relations and permutations between those. It is therefore the exact opposite of an interior knowledge intended as a central and resolutive unity, preceding our experience of another sphere. (Prodi 1974: 18)

There is no privileged access to the interior world (Ryle 1949). In the natural world, everything happens in the light of day or in the shadow of night. Interiority is nothing but the other face of exteriority:

> every distinction [between internal and external, mind and body] is a second step. [...] So, the discourse about knowledge begins from the fact of the interaction between the two domains, which occurs, structurally and inevitably, at every moment as an integral part of the process of knowledge itself: the possibility of setting it apart, to bracket it even just for an instant, is nothing but a myth. (Prodi 1974: 27)

Before examining this argument closer, let us try to detail the explanatory model used by Prodi when giving an account of human knowledge and of the natural history of the I. We are already familiar with the general model: that of the circle. But in this case, it is expedient to refer to a topological morphism of the circle: the Möbius strip.

The Möbius strip has two distinct faces, but it is also a continuous surface. This means that, for a hypothetical organism, walking the surface of the strip starting from any point, it will be possible to return to that same point having traversed its entire surface without taking her own feet off the strip. There is no dualism between two sides of the strip, yet it undoubtedly has two sides. Two sides, one strip surface. Two distinct advantages are gained by employing this figure as a model for human subjectivity, thus replacing Cartesian dualism: (1) The unity of the human animal is not questioned. An explanation of the I grounded on this model is thoroughly naturalistic. (2) This model preserves the constitutive duality of consciousness: always consciousness *of something* and in particular of oneself in self-consciousness. There is both continuity and discontinuity; there is the monism of the human animal and the dualism of consciousness. Both nature and consciousness are preserved. Once again, it is clear why the model of the circle is so dear to Prodi: it allows to avoid both reductionism and dualism.[1]

Before moving on to human consciousness, we should take a closer look to consciousness in general, at a more elementary level. Let us consider the simplest of examples: the encounter of two molecules. Not every encounter is possible, since certain combinations are foreclosed by the physical structure of the molecules themselves. From the point of view of a molecule, this means that certain combinations are "meaningful", while others are not. A combination is meaningful only relatively to another specific molecule. Indeed, to say that "something is meaningful"

---

[1] The first application of the Möbius strip to semiotics can be found in Lo Piparo 1992.

amounts to saying that "something is meaningful" *to* something else (Prodi 1979: 187). It is meaningful because it allows the formation of a larger assemblage:

> an organism has its external correlates to which it is adapted: the things it can exploit. These are meaningful for it. Let us consider these correlates at their highest genericity (from oxygen as a gas necessary for life to oxygen as an object of chemical-physical hypothesis). The adaptation was phylogenetically organized: some things in the environment "become" meaningful for the organism because the organism evolves, thus becoming capable of reading and interpreting them. (Prodi 1979: 187)

At the beginning, as we know, there is a completely natural phylogenetically established complementarity. This is before the emergence of distinction between subject and object, in the realm of relations between *things*: "meaning is not an abstraction or an abstract correspondence, but a relationship between concrete things [...]: it is not a content poured into a vessel, but a fact defined through evolution" (Prodi 1979: 188). Let us now go back to Fig. 6.2, where A and B form the assemblage AB. At the same time, A does not form anything with C, D, or E. To say that A is "interested" in B, yet uninterested in all the other things in its "surroundings", is no concession to anthropomorphism, because in this case "interest" simply refers to the possibility of forming complex structures. A possibility that, in turn, is determined by A's make-up, its material structure. In this sense the "complementarity" between A and B is thoroughly natural. That A is "interested" in B means that A "pays attention"— in a completely natural way—to B while not "paying attention" to C, D, or E:

> there is no organic function (in any organism) that is not buttressed by specific correspondences between molecules, selective processes wherein an interpreter reads signs. These are necessary for life: life is a sequence of interpretations that feed energy to the reader (as a categorial machine, as a reading apparatus oriented towards the world). Interpretation is always necessary. (Prodi 1989: 94)

A living relation entails a selection, among all present things, of those that allow the formation of larger and more articulate assemblages. These things "mean" that a relation is possible. Natural meaning, then, is nothing but a thing's capacity to selectively establish links with other things—where selection means capacity to pay attention to something rather than something else. Finally, bio-semiosis means the ability to paying attention, i.e. to "select". There is no semiosis without this "control" of attention. Such is the fundamental difference between the relations supported by "material logic" and those supported by "categorial logic"—the first kind is "fundamentally nonpreferential", while the second:

> marks the first step towards a preferential reading. [...] In nature occur both situations of preferential reading and of interpretation: these are material states that react selectively with other material states — and only with them — interpreting them operatively. They 'categorize' reality. (Prodi 1982: 40)

There is a crucial nexus between semiosis and attention. We are now in a position to define "consciousness" as this primordial capacity for selectively paying attention to things in the world (Cimatti 2000c). Based on this definition, we can see that there is no life—no semiosis—without a primitive form of consciousness (Emmeche 2004; Baslow 2011).

Returning to our example, A is conscious because when in the presence of B, C, D, and E, it "chooses" B. This kind of definition does not contain any dualist presupposition nor does it entail that A would somehow contain within itself a special entity called "consciousness". On the contrary, "consciousness" means nothing but the ability of selectively form assemblages with other things. In turn, this is not an intentional "mysterious" (McGinn 1993) ability, since it is nothing but the thoroughly non-intentional result of the physical morphology of A and B. This is worth repeating: we can talk of "consciousness" with regard to a thing only if we refer to a selective movement towards other things—a "selecting object" is therefore a "conscious" one:

> [such an object] will come in contact with an indefinite number of other objects, and it will only undergo changes through such contact. Only thanks to this contact it will preserve its coherence: otherwise it will decay (for ex. it will be disassembled to its component elements). (Prodi 1982: 42)

Attention and consciousness have no intentional or "subjective" character, since "categorization" is nothing but a "point of view" on the world (Prodi 1982: 43). Every point of view can discern only a particular fraction of the world, the one that—relatively to that point of view—is meaningful or interesting (here Jakob von Uexküll's influence on Prodi is extremely clear).

This capacity for selective attention towards the world defines the domain of life. And according to this perspective, every form of life possesses it: whether vegetable (Brenner et al. 2006) or animal (Griffin and Speck 2004; Mendelson et al. 2016). It is evident how this is a radical critique—a thoroughly naturalistic critique—of the Cartesian standpoint. Usually the latter is criticized by arguing that non-human animals too have something similar to human consciousness. However, to target Descartes' exclusion of animals from the process of knowledge paradoxically means failing to criticize his dualism: by defending the earthworm's right to have a consciousness one is simply extending Cartesian dualism all the way down to the earthworm. But a more radical critique of Descartes rejects dualism itself. It is not merely a matter of granting consciousness to the earthworm, or to an onion, but rather to deny that the human animal would somehow not be included in the natural world. Dualism is the problem, not consciousness, and with his biosemiotic theory of consciousness, Prodi is launching a direct attack to dualism. In nature, consciousness is not *something* separated from all other material things. For Prodi this is an anti-Cartesian position because it denies consciousness—the consciousness of the dualist paradigm—to the human and not because it extends it to the earthworm.

But if this is the common background of all living (and non living) beings, how does the specific human consciousness emerge? We should begin by clarifying the problem: the issue is not that of searching, within humans, for a special kind of entity that dualism calls "consciousness". Such an entity does not exist. Rather, it is necessary to understand how the development of the human capacity for experiencing the "I" has been possible. Once again, this does not entail the search for a special substance called "consciousness" but rather a peculiar relation that human beings—as well as, it seems, certain animals trained to use a complex form of human communication (Tomasello and Call 2004; Lyn 2017)—have towards themselves. Human consciousness is not a thing within *Homo sapiens*; it is the linguistically mediated means for humans to establish a relation with themselves. To be more precise, human beings *qua* humans *are* nothing but such a relation.

Prodi's most evident inspiration, when tackling this problem, is Vygotsky's historical-social psychology (Cimatti 2000c), a very well-known and popular doctrine in Italy during Prodi's time. We should remember that Prodi was living and working in Bologna, the capital and largest city of the Emilia-Romagna region, well known in Italy as well as abroad for the quality of its kindergartens and elementary schools. The pedagogical model employed in these schools was—and still is—that of historical-social psychology. According to Vygotsky,[2] the psychic development of human beings is not an internal process of growth; on the contrary, every psychological capacity—from sensation and perception to higher cognitive functions—is formed through the relation with the social instruments that children encounter in the community within which they grow up. For Vygotsky, then, nobody is immediately born qua human—rather, one *becomes* a human. In particular, a child of *Homo sapiens* becomes a fully-fledged human only when he or she is able to use upon him- or herself the external social instruments offered by the social environment where his or her development took place. The most important of these is speech:

> the transition from the biological to the social way of development constitutes the central link in this process of development, the cardinal turning point in the history of child behaviour. This road — passing through another person — proves to be the central highway of development of practical intellect, as demonstrated by our experiments. Speech here plays a role of primary importance. (Vygotsky and Luria 1994: 116)

In particular, the child—at a certain stage of his or her cognitive development—begins to use upon himself or herself those forms of social behaviour, specifically linguistic ones that were previously used only towards others. For example, at first words would only be used in order to ask something to someone, a way for the child to "control" the behaviour of others. But what happens when there is nobody to ask for help to? It is in this circumstance that, according to Vygotsky, a radical change takes place: the young human tells himself or herself what to do. That is to say, he or she starts to actively control her/his own behaviour. The child's self-consciousness is, for Vygotsky, nothing but this linguistic capacity for self-control:

> [t]he greatest change in child development occurs when this socialized speech, previously addressed to the adult, if turned to himself, when, instead of appealing to the experimentalist with a plan for the solution of the problem, the child appeals to himself. In this latter case the speech, participating in the solution, from an inter-psychological category, now becomes an intra-psychological function. The child applies to itself the method of behaviour that it previously applied to another, thus organizing its own behaviour according to a social type.

---

[2] Nowadays it is becoming less and less certain that the texts traditionally attributed to Vygotsky were actually written by him alone (Yasnitsky, Van der Veer 2016). However, the ideas that one can find in Vygotsky and Luria's *Tool and symbol in child development* are very similar to those found in the books that were published under Vygotsky's name when he was still alive. For this reason, I believe that *Tool and symbol in child development* can still be considered a reliable source for Vygotsky's own ideas.

> The source of intelligent action and control over his own behaviour in the solution of a complex practical problem is, consequently, not an invention of some purely logical act, but the application of a social attitude to itself, the transfer of a social form of behaviour into its own psychological organization. (Vygotsky and Luria 1994:119)

What was once an intra-psychological behaviour, taking place between people, now becomes and inter-psychological one, within the child's psyche. Properly speaking, then, the conscious psyche of the child is nothing but this capacity to apply to oneself a social instrument. What before was said to others is now said to oneself: human consciousness is nothing but this capacity to speak to oneself. Through inner speech (or "verbal thought"), the young human becomes able to pay attention to his or her own behaviour. This is the reason why consciousness is not a thing, since it is nothing but this capacity for self-attention. Therefore, the child does not speak to him- or her-self as if two distinct entities resided within: the one that speaks—the mind—and the object of the speech, the body. The child's consciousness is the very act of speaking to him- or herself: it is a reflexive use of language. More precisely still, the intra-psycho-logical use of language allows the child the pay attention to his or her own behaviour. The primary function of "verbal thought"—i.e. an interior form of social speech—is the voluntary focusing of one's attention. The "will" is nothing but the control of one's behaviour instantiated by the "verbal thought" (be it explicit or implicit):

> [f]rom the first steps of the child's development, the word intrudes into the child's percep-tion, singling out separate elements overcoming the natural structure of the sensory field and, as it were, forming new (artificially introduced and mobile) structural centres. Speech does not merely accompany the child's perception, from the very first it begins to take an active part in it: the child begins to perceive the world not only through its eyes, but also through its speech, and it is in this process that we find an essential point in the development of the child's perception. (Vygotsky and Luria 1994: 125)

We once again encounter the theme of the control of attention, a central issue in Prodi's thought: "with the help of the indicative function of words, noted above, the child begins to master its attention, creating new structural centres of the perceived situation" (Vygotsky and Luria 1994: 132). Human consciousness, unlike that of animals or plants, coincides with the inner capacity of directing one's attention. Specifically, the child's behaviour becomes truly "his" or "her" behaviour only once the ability of controlling it through "verbal thought" is acquired. The distinction between the body and the mind, that is, between the action and the conscious con-trol of such an action (this is the consciousness) is nothing but the "by-product" of the self-reflexive use of language. Building upon Vygotsky's theories, Prodi thus manages to deliver a plausible account as to how a young human being—a primate very similar to anthropomorphic monkeys (Gagneux and Varki: 2001)—can develop the control of his or her attention first and acquire consciousness later. This is a fully naturalistic (non-dualist) account, articulated within a framework of continuity between the animal and plant world (the domain ruled by "categorial logic"), which is a complication of the natural world (that of "material logic").

Going back to Fig. 8.1, we now fully understand the expediency of the Möbius strip model. According to the latter, consciousness appears whenever a speaker using language reflexively (Dennett 1991) institutes a momentary split between

Fig. 8.1 The Möbius strip
as a model for human
self-knowledge

inside—a speaking voice—and outside, the bodily behaviour that the voice is describing. This is not an absolute separation, and it is certainly not a dualism: on the contrary, it is a distinction that is instituted by the act of reflexive speech and lasts only as long as the latter. As Benveniste (a linguist whose work Prodi was well acquainted with) put it:

> [i]t is in and through language that man constitutes himself as a subject, because language alone establishes the concept of "ego" in reality, in its reality which is that of the being. The "subjectivity" [...] is the capacity of the speaker to posit himself as 'subject'. (Benveniste 1971: 224)

Human consciousness—i.e. the "I"—is nothing but a "collateral effect" of the reflexive use of language. In particular, consciousness is the attention that the speaker turns to him- or herself through the employment of language:

> [w]hat then is the reality to which *I* or *You* refers? It is solely a "reality of discourse", and this is a very strange thing. I cannot be defined except in terms of "locution", not in terms of objects as a nominal sign is. I signifies "the person who is uttering the present instance of the discourse containing I". (Benveniste 1971: 218)

There is nothing left of Cartesian dualism. The Möbius strip shows us how there is only the natural world but that this world can fold upon itself without breaking its continuity: human consciousness is the natural fabric of the world folding upon itself. This is only possible for the linguistic animal par excellence, the human animal, the only animal who has made of language its own environment and who has adapted itself to it and constructed itself around it:

> [f]or every thing that is recognized and linked to spoken word there needs to be the activation of a sequence of specialized circuits, one after the other, with different filters and at different levels. When spoken word enters into a logical play —used in a context of logical operations — a system of circuits is linked to other systems. This has to be a unitary process, because step by step the cerebral system rotates as if anchored to a central joint, and the entire system works for that point, as if there was no constant centre and every point could function as centre. This variable condensation onto a single point is *attention*, i.e. the process of focusing the entire machinery that, for a given process, puts itself at the service of one of its parts (an object, a word, a person, a logical process...). [...] Therefore, if we interpret reality [...] through a given hypothesis, then it is as if we were building a network with certain connections (materially instantiated by a series of activated circuits [...]) and this network would then "filter" the reality it delivers to us, catching those fish/objects that are the right size and shape to be caught by such a net. (Prodi 1987b: 53)

Prodi's image is both simple and efficacious: the words of a language allow us to catch "fish/objects", i.e. the entities we can turn our attention to. The point is that these fish can be caught only because they let themselves be caught. This means that the network of language is built on the world; that it, language is a phylogenetic adaptation to the "fish/objects". Language can say the world only because *it is* made of world: language is a part of the world that talks about another part of the world. This is a point that Prodi shares with Jakob von Uexküll:

> Since Uexküll believed that this activity of the mind consists in the reception and decoding of signs, the mind-in the final analysis-is an organ created by nature to perceive nature. Nature may be compared to a composer who listens to his own works played on an instrument of his own construction. This results in a strangely reciprocal relationship between nature, which has created man, and man, who not only in his art and science, but also in his experiential universe, has created nature. (Thure von Uexküll 1987: 149)

This is the kind of predicament illustrated by the Möbius strip: language is the world folded upon itself, a questioning about the world starting from the world itself—and such a questioning is nothing but human consciousness. The latter, therefore, is not a starting point: not of semiosis nor of language. As Prodi writes:

> once examined, the facts of "consciousness" appear grounded on natural bases, unconscious and automatic, and can only exist thanks to them: consciousness, then, is just the tip of the iceberg, and if we intend to explain anything at all (i.e. put it in connection with something else) we need to invoke the submerged part, that which allows the tip to surface in the first place. (Prodi 1977: 18)

Chapter 9
Breaking the Circle

The evolution of language coincides with the evolution of consciousness, which is in its very nature linguistic. The broadening of the field of inquiry is also a broadening of language.

(Prodi 1974: 226)

Abstract Prodi explains semiosic phenomena by means of the model of the circle. This model allows to account for a very important feature of the natural world: in life phenomena, there is no a one-way causal arrow, from the lenvironment to the living organism and vice versa. According to Prodi in the lifeworld, there is a two-way relation between cause and effect; the genotype causes the phenotype, but the latter in turn modifies the evolutionary environment, thus ultimately applying an indirect causal pressure back on the genome itself. In this way, it is possible to account for the dynamism of the world of life. The biological world, the world of semiosis, is not closed off, but it continuously expands and develops. The circle of life is best conceived as a spiral. The transition from the circle to the spiral is necessary in order to account—in a biological register—in particular for the evolution and the mutation of human language and culture.

Keywords Language and world · Language and knowledge · Model of the circle and of the spiral

According to Prodi, human language is a biological phenomenon. However, he also considers culture to be a biological phenomenon. A phenomenon is defined as cultural when it is not determined by genomic necessity only, when it allows a more or less degree of freedom and choice. But Prodi also considers complementarity (see Fig. 6.2) to be a cultural phenomenon. As we have seen, everything in the world is a "reader" of its surroundings; but "to read" means to *interpret* (since reading is not a simple decoding; to read means to select what is relevant and discard what is not relevant). This means that even the most elementary of natural phenomena are somehow cultural, i.e. not completely predetermined: "in nature complementarity is

© Springer Nature Switzerland AG 2018
F. Cimatti, *A Biosemiotic Ontology*, Biosemiotics 18,
https://doi.org/10.1007/978-3-319-97903-8_9

above all a reading, an interpretation, that is to say an exchange subordinated to a reciprocal meaningfulness" (Prodi 1977: 26). As we know, something can be meaningful for a particular thing and not for another thing: meaningfulness implies selection and choice, even though this is an unconscious and non-intentional choice. Going back to our example of two things, A and B:

> the process of encounter is therefore a reading or interpretation of reality, operated by A and B according to their constitutive modules. Therefore, A's and B's reading of reality is a survey of the environment, discarding irrelevant things and selecting those that are meaningful. Through the exploratory process that terminates with its contact with B, A evaluates reality, since it is prompted to change only when it finds a complementary and meaningful reality. (Prodi 1977: 26)

Nature is both within culture and within language, but this also means that there is something cultural (reading and interpretation) within nature. It is necessary to recognize the error of a nature-culture dualism and rather understand that culture is natural and that—something which will surprise naïve materialists—nature, and human nature in particular, is somehow cultural:

> [m]an is constitutively characterised by communication through systems of exchange. It differentiates itself from other organisms because it is capable of acting through the mediation of an abstraction [...]. This is [...] a radically new way of interacting with nature [...]. Man is genetically gifted with linguistic competence but languages are not a mere fuel for the functioning of the logical machine. Languages have two sides to them: on the one hand, they are used to communicate, like an instrument. They are useful because they have been designed for this purpose. On the other they also have a formative function, expressed when the individual begins to apply his competence [...]. It is as if the fuel would contribute to the construction of the machine it fuels. (Prodi 1987b: 196)

Dualism, in all of its forms, considers biological phenomena as guided by principles that are wholly different from those of culture and of history. Prodi's proposal takes us, in one fell swoop, beyond this quaint (but ever-returning, see Lavazza and Robinson 2014) opposition between nature and artifice, between biology and culture. The expressive power of human language does not depend on an arbitrary bestowing of meaning to a signal; it is rather the final manifestation (ignoring its own origins) of a natural meaningfulness whose primordial exemplification is to be found in the most elementary biological interactions:

> [l]anguage cannot but be a phenotypical or functional expression of an underlying "language skill", that makes itself manifests in the very first forms of biological organization and that gets increasingly complex after that. Language is necessarily grounded upon a — wholly pre-linguistic — historical connective layer of reciprocal meaningfulness that, at a given moment, is *also* able to constitute a language or, alternatively, to evolve into a language. The ever more complex logic of connection of meaning constitutes the deep structure of language: as a function it is somehow unique to man, but it can be explained only where understood through its biological and genetic roots. (Prodi 1983b: 189)

In this way nature itself becomes a kind of endless and infinitely complex "translation network": "every chapter of molecular biology is the exemplification of how meaning progresses in nature: through the elaboration of structures and translation chains (for example, consider protein synthesis, the transmission of genes from two individuals along the germline, muscular contraction, or the transmission of

Fig. 9.1 Continuity
between natural semiosis
and human language

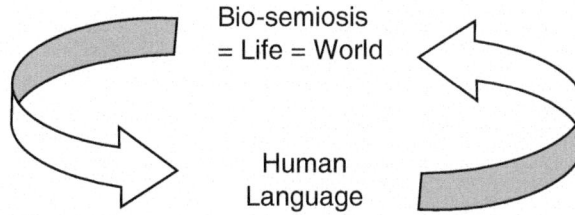

Bio-semiosis
= Life = World

Human
Language

impulses)" (Prodi 1983b: 192). For Prodi, the key notion to understand the world of semiosis is *translation*: "reading never takes place simply at one stage, it is never mono-categorial [...]: organisms converge towards the outside in complex readings. This convergence towards the outside coordinates and unifies the large number of mono-categorial internal readings" (Prodi 1983b: 190).

But if nature is an immensely complex translation network, an unending sequence of translations/transformations of objects into other—more complex and stratified—objects, a problem seems to emerge, which indeed afflicts all philosophies based on this schema: a problem intrinsic to the model of the circle. A circle is closed: there is, by definition, no escape from the circle. But life is change: that is to say that life is a continuous breaking of the circle. It is then necessary to understand how the circle can be broken and be transformed into a spiral. In the meantime, let us go back to the circle (Fig. 9.1).

This diagram illustrates the continuous line linking language to natural semiosis and, therefore, to the world:

> if we move from generic biological situations to language, the connective function that is proper to it appears as grounded on enormously complex and interconnected translation chains [...] yet always respecting the transmission of the meaning of things to the reader, and then the reader's answer to this meaning, that is to say back again to the things. For language too, it is possible to state that, ultimately, things let themselves be read or spoken of. (Prodi 1983b: 192)

Language can speak about the world because it is nothing but the last, and most complex, manifestation of the primordial semiotic capacity of life itself, i.e. semiosis:

> [when] the identifications of meaning are so complex that they produce a separation between sign and referent [...] it means that the chains between them are extremely long and complex, and that they can be only described through their emergent phenotypical aspects — those of an already-given language, to be analysed in its factual communicative features. Yet, biologically speaking, this situation must be based upon long translation chains with well-determined steps, capable of conditioning whether or not a selective reaction takes place. Linguistic reading is not a miniaturization of the real, and words are very different entities than the things they describe: but between a reading of the world and the world itself there is a semiotic link that derives from the composition of conditions of elementary proto-semiotics — in the form of extremely complex chains — both for what pertains to the structures meant for linguistic exchange, memorization, and learning, and for what pertains to the mediation globally represented by the whole of language and culture. (Prodi 1983b: 192)

We can speak about the world because—and herein lies the strength of the bio-logical model of the circle—the world lets itself be spoken of through "our" lan-guage (in fact it is the language of nature), since "we do not attempt to explain things through our mind, but our mind through things" (Prodi 1987b:134). But Fig. 9.1 also shows another vector that starting from language and moving towards biological semiosis, i.e. the world. As we have already seen, Prodi does not want to merely root language into the world, but rather—faithful Darwinian that he was—he wants to highlight the peculiarity of human language vis-à-vis all other forms of semiosis. The diagram above seems to paradoxically suggest that—somehow, and once a certain threshold of complexity is crossed—language could become its own cause. This is an extremely important point, one only recently re-examined in the context of the debate on the evolutionary history of language (Deacon 1997). Once again, this is a theme deeply rooted in Jakob von Uexküll *Theory of Meaning*:

> Each subject's symbol is at the same time a meaningful theme for the structure of the sub-ject's body. The body that houses the subject on the one hand produces the symbols that populate the surrounding garden and is, on the other hand, the product of these very same symbols that are the meaningful themes in constructing it. The sun owes its shine and its form high up in the sky that extends over the garden to the eye, as the window of the body that houses ourself. At the same time, the sun is the theme guiding the construction of the window. This principle applies to both animal and man; the same factor of nature manifests itself in both cases. (Jakob von Uexküll 1987: 113)

The origin of human language should not be sought in a neutral external reality, as it is frequently assumed by those who think that the main problem of a natural theory of language is the so-called symbol grounding problem (Harnad 1990); such a non-biological reality constitutes the common background for all forms of animal communication, and for this reason, it cannot represent the *specific* environment wherein the origins of human language should be sought. This element, Prodi argues, needs to be sought into language itself: the latter—once crossed a cer-tain threshold of complexity—has determined the conditions for the development of ever more complex linguistic forms. Consider the case of the wings of birds: it is very probable that the first rudimentary forms of wings developed in dinosaurs did not serve the purpose of flying but simply functioned as highly visible surfaces dis-played during mating rituals (Zelenitsky et al. 2012). But in order to be efficacious, these surfaces needed to be sufficiently light and large; at a certain point—as an unexpected collateral effect—these allowed the animals who had them to achieve brief flights. Now, the selective push does not only (or not simply) derive from a reproductive advantage but from flight too: the wing's structural complexity increases because evolution develops forms that are better adapted to this new and completely unexpected purpose. Flight, now the implicit purpose of the wing, selects forms that are ever more suitable for flying. The evolutionary push towards flight becomes the primary "engine" for the development of increasingly sophisti-cated wings. The evolutionary scenario that frames the development of the wing is now represented by the wing itself that, in this sense, becomes its own cause—the cause of ever more complex kinds of wings. Prodi argues that the evolutionary

The spiral becomes progressively wider and produces ever more extended areas of "darkness". Knowledge does not reduce the amount of ignorance: on the contrary, it increases it. This is an implicit conclusion of Prodi's reasoning, according to which knowledge is always a "process" (Prodi 1974: 26), always relational and hypothetical: "the starting point is the contact with the terms of the process; not simply to accept them, but to experiment with them" (Prodi 1974: 27). For Prodi—like Peirce, from whose system Prodi's biosemiotics largely derives—there is no intuitive and unmediated knowledge:

> [m]oreover, we know of no power by which an intuition could be known. For, as the cognition is beginning, and therefore in a state of change, at only the first instant would it be intuition. And, therefore, the apprehension of it must take place in no time and be an event occupying no time. Besides, all the cognitive faculties we know of are relative, and consequently their products are relations. But the cognition of a relation is determined by previous cognitions. No cognition not determined by a previous cognition, then, can be known. It does not exist, then, first, because it is absolutely incognizable, and second, because a cognition only exists so far as it is known. (Peirce CP: 5.262)

The lack of an absolute starting point for knowledge also entails that the epistemic process is endless. Every new piece of knowledge needs to be understood, and therefore an "interpreter" capable of "explaining" it is necessary. A hypothetical "final" knowledge, if knowledge at all, would still need to be understood, and it would therefore require an "interpreter"—if so, it cannot be *final* at all. Hence it is language itself that continuously produces new "darkness". This point needs stressing, since it is what prevents the semiosic circle from becoming sterile, a vicious circle. According to the model we have thus far reconstructed, world and semiosis are coextensive. But does this mean that language completely covers the whole extension of the world? Does language fully "swallow" the world? Does biologizing language perhaps mean to eliminate once and for all every non-linguistic horizon? In Fig. 9.1 and Fig. 9.2 there is a first answer to this question: human language is located *within* an indefinitely more extended space—that of "darkness"—which, by definition, *exceeds* its logical resources, a space that is, and cannot but be, located *beyond* language. In fact, the passage from the circle to the spiral represents the passage from a (relatively) closed environment to an open one; the "darkness" is such an environment which is full of epistemic "problems". The "darkness" is what is produced when the world is read through language. The "darkness" is not simply the absence of world/language; on the contrary, it is the region from which language as a whole assumes its meaning: only insofar as it produces "darkness" can language have meaning. The presence of "darkness" means that language is not everything and that there exists something which is not language. Not only language cannot exhaust the world, but the signifying power of language seems to derive precisely from this exceeding of language by the world. Wittgenstein illustrates this point with exemplary clarity:

> 6.41 The sense of the world must lie outside the world. In the world everything is as it is and happens as it does happen. In it there is no value—and if there were, it would be of no value. If there is a value which is of value, it must lie outside all happening and being-so. For all happening and being-so is accidental. What makes it non-accidental cannot lie in the world, for otherwise this would again be accidental. It must lie outside the world (Wittgenstein 1922: 87–88)

Fig. 9.2 The evolutionary
spiral of human language

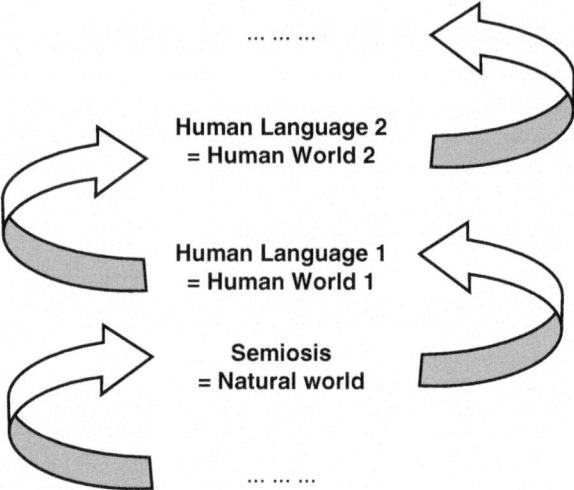

This space beyond language, which can only be thought from *within* language, is precisely the space that justifies and grounds language itself, since "[t]he sense of the world must lie outside the world". Indeed, language is not one with the world, as in an idealist construal (if someone has ever really held such a thesis) according to which language creates the world, being its very substance. The evolutionary identity of language and world that Prodi insists upon is not a form of absolute idealism. Language is inscribed within the world, derives from the world, and is a part of the world (and Fig. 6.2 clearly illustrates this point). The meaning-bearing relation *AB* can be established only because other things—*C*, *D*, and *E*—remain indifferent (this space of indifference is the "darkness"). The non-semiosic world is always wider than the semiosic one. Semiosis means selection, in the context of "categorial logic", and hypothesis, in "propositional logic":

> [t]he formulation of hypothesis only develops in man, and characterizes his epistemic process. It allows to imagine possible states of reality, based upon of data gathered according to the same logic (an extension of material logic) that they followed while being assembled. The individual de-assembles and recombines these data, finding a best fit with reality [...]. The result are constructions that can be compared with reality — possessing a certain degree of analogical correspondence with it — through processes of translation: in this sense internal representations are also signs, and can be translated into communicable signs. These constructions can be arranged not only according to reality, but also to other strata, completely detached from reality, to be taken as reference points. (Prodi 1977: 132)

Prodi is here telling us that hypotheses are not the result of an inexplicable creative act. As we have already seen, when "a sensorial entity, instead of following its relatively fixed groove, enters into the sea of language" (Prodi 1987b: 217), this stimulus enters into an indefinitely extended combinatorial space, within which hypotheses can be formulated. The hypothesis is nothing but this reverberating of the incoming stimulus between the nodes of the internal network of language. Thus, the hypothesis—as an experimentation of possible combinations of data from the internal deposit—breaks the closure of the environment, making it possible for the consciousness of boundaries to emerge. As a matter of fact, the discovery of "darkness" means that the world does no coincide with language: in this sense, we can say that hypothesis—at the origin of aesthetic experience and of the sacred—is an intermediate state between "incapacity and desire" (Prodi 1987b: 217) and represents the most proper human experience, pertaining to the "problem of the boundaries" (Prodi 1987b: 217) of our world. Language as hypothesis institutes the separation between environment (the one in Fig. 4.2, that is to say the closed space of the animal's "functional circle") and human world. The human world is composed of both the linguistic environment and the "darkness"—the world is real just because language is *not* everything:

> [i]t is precisely the hermeneutic weakness of language that allows the world to be a "world". Only when the word shows its powerlessness to decode the non-linguistic can the latter domain present itself as a context that cannot be transcend — thus constituting a *world*. Somewhat paradoxically, we could say that *the world is linguistically constituted by that which, within language, betrays the incomplete or limited nature of language itself with respect to the world*. Ultimately, the fact that men have a "world" (wherein merging is always imperfect, conflicts are always unresolved, and adaptations are partial and precarious) instead of an "environment" (into which organisms as irrevocably integrated, as if immersed into an amniotic fluid), is explained by the limits of language rather than by its representative power. (Garroni 1986: 263)

We can once again return to Wittgenstein's famous proposition "[t]he sense of the world must lie outside the world". This means that this gap between internal and external, between language and nonlanguage, determines the conditions for another paradox: that proper of those who, from *within* the world/language, aim to grasp that which lies outside the world—an ambition accompanied by the clear and tragic awareness that this step towards the outside will forever remain logically impossible. From this point of view, *we are* this paradox: our very essence as human animals is distilled in this condition of suspension, this substantial "perplexity" (Prodi 1987b: 217)—our being "radically dark to ourselves" (Prodi 1987b: 216). This is not a paralyzing paradox though: on the contrary, it represents the fundamental push that, operating from within the world/language, propels the continuous attempt to extend and transgress its borders. This is the opposition that the Italian philosopher Emilio Garroni (one of the most important Italian philosophers active while Prodi was developing his own biosemiotic model) examined in his *Senso e paradosso*, pertaining to an experience that, on the one hand, "run[s] against the boundaries of language" (to use Wittgenstein's phrase [see Chap. 10]) while, on the other, can be meaningful only insofar as it runs against such boundaries. Human language can signify something only because it cannot signify *everything*:

> [a]s I see it, the irrational is not defined by its opposition to the rational, but it rather indicates the whole ensemble of things, only a very limited part of which we can rationalize. It produces knowledge and, eventually, consciousness. The irrational is therefore a large container: there is always more that can be said about it, but it still remains a container in principle unobservable for us from the outside, always imposing limits on us which we see as boundaries and foreclosures. Beyond these borders, there are unrecognizable lands. (Prodi 1987b: 213)

Chapter 10
Language and Ethics

This is arguably the point: the artificial is, perhaps, a simple extension of the natural. It may be that the natural too has its own artificial mechanics, as if it had been invented by someone, or if it has been inventing itself during its process of self-construction.

(Prodi 1987b: 33)

Abstract The biological ground of ethics, according to Prodi, should not be sought in feelings of empathy or altruism. On the contrary, human ethics is profoundly "unnatural", precisely because it is unbound from any genetic principle. As a matter of fact, if there was such a "natural" morality, there would be no free choice, since human behaviour couldn't but conform to these "natural" norms. But with no freedom of choice, to speak of ethics becomes meaningless. According to Prodi, ethics can exist because language—that is, hypothesis and choice—exists. The natural ground of ethics is our faculty to use a language.

Keywords Ethics · Empathy · Altruism · Freedom

Knowledge, since its biological beginning, has always had a moral character.[1] Prodi writes that "the beginning of knowledge is [...] a moral moment; a fleeting priority is assigned to that which lies outside" (Prodi 1987b: 71). Morality emerges, in nature, when the metabolization of the external object is suspended, thus "allowing" that object to exist independently from the use one can make of it: it is this suspension of the consumption of the object that opens the door to the possibility of morality. But this is not the customary way to approach questions of morality: it is unusual to start with nature, adopting a bottom-up approach; rather, following a top-down strategy, the starting point is moral consciousness. It is in fact believed that morality,

---

[1]Although Prodi talks mostly of "morality", he really is concerned with *ethics*, as the set of all biological and linguistic conditions necessary for a human being in order to have ethical experiences.

© Springer Nature Switzerland AG 2018   99
F. Cimatti, *A Biosemiotic Ontology*, Biosemiotics 18,
https://doi.org/10.1007/978-3-319-97903-8_10

on the one hand, implies a distinction (yet another form of dualism) between a norm that establishes what is good and what is bad and, on the other, that it also presupposes the ability to choose between the two. A further assumption lies beneath these presuppositions: that morality implies the presence of a subject, the only entity able to choose freely. This kind of perspective seems irreconcilable with Prodi's naturalistic approach. What the usual way of considering morality wants to preserve it is the autonomy of the moral subject. Otherwise there is always the danger of incurring into Moore's "naturalistic fallacy", confounding the natural and the ethical: that something is "natural" does not mean that is "right", and therefore "good is not to be considered a natural object" (Moore 1959: 14). The "good" pertains to the world of human relations, mediated by language and cultural traditions, and it is not a thing of the world, like a pear or a fish.

Prodi, as usual, defends a very original position. On the one hand, he joins the ranks of those who attempt to naturalize morality (Singer 1981; de Waal 1996; Boniolo and De Anna 2006) but, on the other, he seeks a rather different "naturalization" than most (Hauser 2006). Take the example of altruism, a very common behaviour among non-human animals (Fehr and Fischbacher 2003; Schino and Aureli 2010). Here's one of the many possible accounts of this behaviour:

> [t]he story of a female western lowland gorilla named Binti Jua, Swahili for "daughter of sunshine," who lived in the Brookfield Zoo in Illinois. One summer day in 1996, a three-year-old boy climbed the wall of the gorilla enclosure at Brookfield and fell twenty feet onto the concrete floor below. As spectators gaped and the boy's mother screamed in terror, Binti Jua approached the unconscious boy. She reached down and gently lifted him, cradling him in her arms while her own infant, Koola, clung to her back. Growling warnings at the other gorillas who tried to get close, Binti Jua carried the boy safely to an access gate and the waiting zoo staff. (Bekoff and Pierce 2009: 1)

Binti Jua, according to human customs, behaved in an altruistic and caring way towards the little human who fell in the gorilla enclosure. But that is from *our* way of interpreting the situation: we could presume that some empathic acknowledgement was triggered in the gorilla (O'Connell 1995; Presto and de Waal 2008) that led her to protect the young human. This would be a natural behaviour, with a clear evolutionary explanation. But its naturalness does not entail its moral "rightness", and most importantly it does not even mean that it can be considered a "moral" behaviour to begin with. A human being, behaving in the same way in the same predicament, could have done so without any feeling of empathy towards the child and simply out of a feeling of duty. A human action is not moral because it is motivated by empathy (or immoral because motivated by antipathy) but because it is *considered* "moral" by the community to which the actor belongs. Morality pertains to the rules of society, not to the innate endowement of the individuals. Empathy does not explain anything about human morality: the latter is properly human precisely because it is independent from empathy or antipathy. How could we "praise" an action based on empathy if this was actually an innate behaviour? There would be no merit in being empathic, no more than there is in having two lungs and a bladder. On the contrary, a moral action is all the more "praiseworthy" when it manages to overcome an immediate and "natural" feeling of antipathy.

If to "naturalize" human morality means to reduce it to the behaviour of non-human animals, then this project can give no insight into what makes a *human* action a *moral* action. The kind of naturalization pursued by Prodi, on the contrary, looks for an "explanation" of human moral behaviour in its own species-specific nature. As we have seen, for Prodi, this means to look into linguistic ability: "the root of our humanity is to be sought in the formation of our linguistic competence" (Prodi 1987b: 49). When looking for the reasons for our moral behaviour, we need to analyse human nature—that is to say, language. Let us begin by examining which elements of our language make human moral behaviour "natural", starting with the link between language and consciousness. Knowledge entails the search for explanations, and this is, first and foremost, a linguistic act: "from this point of view science is a 'natural' activity that develops the fundamental logical and linguistic competence of our species" (Prodi 1987b: 82). In order to explain something, it is necessary to have—at least implicitly—a *theory* assigning a shared meaning to the observed facts. And a theory also implies hypotheses, therefore a mechanism that allows their formulation and rectification, should they fail to explain the facts (it is therefore necessary to be able to *negate* something, not just to affirm). Are these operations cognitively accessible to a non-human animal, an animal that can communicate through his natural language but seems incapable of thinking through it?

> [t]he hunter sees the tracks on the ground, and links them to the animal. He knows that the footprints didn't appear by themselves but were left by someone — the animal he's tracking — who walked that way. [...] We could object that some animals are very skilled hunters. [...] But they still demonstrate the difference there is from man's predicament, although they do start from a common ground, i.e. the fact that in nature are already present, before man, astute and laborious forms of knowledge. [...] Having acknowledged this continuity [...] we should immediately stress a profound difference [...]. The hunter can teach his methods to an apprentice. [...] These function as general rules, applicable to different kinds of prey, while the animal possesses only a rigid ability, relative to the one kind of prey it is predisposed to hunt [...]: the animal does not have a *theoretical understanding* of the hunt. The hunter does, he has a *mental* picture [...] that allows him to link together different elements [...]. The hunter can make predictions or formulate [...] *hypotheses*. (Prodi 1987b: 12–13)

In order to *think* reality, one needs a theory—even an implicit one—that is to say a set of assertions, negations, inferences, and so on that constitute a complex network within which a single fact, observed or imagined, can be placed. It is only within this general framework, Prodi argues, that the moral problem can be posed, where the other object is more than a silent and passive terminus for one's actions (I can perceive it, I can hunt it, I can eat it, or I can ignore it), and rather it is considered as another entity with its own autonomy and projects. In order for a genuinely moral gaze upon the world to exist, it is necessary—as a preliminary condition—that a theoretical knowledge of the world be possible. What does an animal need to produce a theory? Prodi argues that:

> [a] theory exists thanks to a faculty to formulate theories [...] Essentially, this competence is to be identified with linguistic competence allowing, through a unique mechanism, the formation of innumerable sentences capable of facing a large number of different and unique situations. (Prodi 1987b: 14)

The foundation of human morality, then, is to be sought in language, because it is only through language that it becomes possible to achieve that *cognitive distance* from things that allows the emergence of a moral attitude towards the world and towards others. There is morality where there is the possibility—logical possibility, before factual—of deferring the moment of action and of its consummation and therefore to safeguard the autonomy of the object on whom the action is directed.

If biology means relation and complementarity, morality emerges from biology when the "translation chain" that links *A* and *B* (Fig. 6.2) is sufficiently lengthened (or loosened) that *B* can begin to be considered in itself, *independently* from *A*, *as if* there was no longer a relation of continuity between the two, no matter how indefinitely extended:

> it is necessary to note that this discontinuity [...] represented by human knowledge is neither a split nor an opposition: it is simply a greater length of the chain of knowledge, leaving untouched the adherence to the real as well as the historical-phylogenetic solidarity with the world. (Prodi 1987b: 42)

As we have seen, morality emerges in nature when *A* lets *B* exist without consuming it: that is to say, when *A* does not immediately "eat" *B*. In this sense, there is a very strict link between moral behaviour and negation. Turning a common assumption on its head—the one that considers empathy as the biological ground of morality—it is *negation* that allows *B* to survive to the interaction with *A*: indeed "negation is unknown to categoriality" (Prodi 1982: 131), i.e. to the cognitive and sensible world of vegetables and non-human animals, the world of nonlinguistic semiosis (Horn 1989). The ethical value of negation is crucial: it makes it possible to distance oneself from the world as presented, to imagine a different world (formulating hypotheses) and stop our own actions—this is the *natural* origin of ethics. Indeed "categorial operations are invariably positive, since they apply to something meaningful *qua* really existent" (Prodi 1982: 131). On the contrary "only with the appearance of a system of knowledge equipped with propositional characteristics does negation become a meaningful operation, serving the purpose of hypothetically and operationally describing the real" (Prodi 1982: 131). Negation, then, is the logical precondition of hypothesis; a hypothetical situation can only be imagined by negating the one we are currently presented with: "the problem of the existence of a thing, even when this is not directly observed, can only be given within a global frame of analogies and propositional operations. It is only there that the question of whether or not a certain hypothesis regarding reality — or more generally regarding real or putatively real facts — acquires a meaning" (Prodi 1982: 132). Let us try now to assemble all the fundamental natural preconditions of morality according to Prodi:

[i] human knowledge [...] emerged from the "objectification" of things. Now, things are not consumed or eaten at will (the targeted destruction that characterizes all of categorial logic). They now are preserved [...]. Even if for just an instant [...] reality becomes more important than my need to consume it. I spare it, and refrain from eating it. I recognize it as a point of convergence of an abstract system of exchange. I therefore suspend my destruction. This is not a collateral effect of knowledge, but rather an integral part of it. There would be no man

without such a moral attitude. [...] [T]he most real "thing" for a man is, in any event, another man. Here the other is protected from consumption, from immediate utility, and is 'pulled into discourse' and 'made part of the linguistic fabric'. [...] Therefore, the preservation of reality (through the objectification that constitutes the beginning of human knowledge) is the first pillar of morality.

[ii] [N]obody speaks alone. Man built himself upon the other. Man, that is, is constitutionally social in a far more concrete sense than all other animals that are organized in societies. Man, built upon discourse, has interiorized the other. [...] Others have selected, in us, the means to communicate with us. We carry the network of all the external agents we have been shaped by *within ourselves*, that is to say, we reproduce this network (Prodi 1987b: 166–167).

[iii] The presence of a field, [...] of a theoretical infrastructure — represented by its various linguistic forms and by its codes — is constitutive of man's proper kind of knowledge. [...] The "self" is cast within this abstract field: it is here that we find the single individual. He thus arrives to self-consciousness, i.e. he can see himself as an operational unit within this field. The identification of oneself amounts to the identification of this "object I" in one's area of belonging. [...] This is not a metaphysical event — identifying a dimensionless point — but a time-bound singular operation, all the more efficient (leading to a more precise consciousness of the self) the wider the history and the geography of the objective field are in the representation of the individual — and the more complete is his language. [...] The fundamental logical-moral operation of logical (abstract) substitution with the other can only take place in the logical-linguistic field, and therefore in the reality that it mediates. Only in this field I can see another me in the other, and can also look at the world through his eyes. (Prodi 1987b: 45–46)

The first condition institutes the logical space necessary for the emergence of moral objects, that is to say entities endowed—at least potentially—with rights. The first and most important of these is the right to have rights, the right not to be immediately eaten and consumed. It is important to note that this condition, in turn, presupposes the existence of complex objects, those "federations of readers" we have already encountered: before morality, but continuously with it, there is always biology. The second condition defines that particular complex object that is the human organism, the subjective I: this constitutes itself—through language—through others and onto others. Speaking with other autonomous agents (the result of the first condition), the knowing object discovers itself as an "object I", since it discovers its substantial relation with all the other "I"s. This is an identity relation because it is an "I" like everyone else, defined by the logical infrastructure of language, and a relation of difference because every "I" presupposes and makes possible a correlative "you" (Benveniste 1971). It is only within language—and within its logical apparatus—that the "I" *qua* self-consciousness can emerge, at once an abstract and a concrete object: abstract because it is given by the entry of the individual into the logical network of language (common to everyone and therefore tending towards uniformity, turning different things into the same); concrete, because every entry into language is always a singular—spatially and temporally defined—event, an *uttering* "I" here and now. Prodi explains that:

the individual enunciates some markers of differentiation, i.e. propositions. [...] These, constructed from the impersonality of the codex, are veritable "ultra-distinctive" traits, that is to say that they do not simply mark the individual, but also that particular state that characterizes it, in a specific instant, within its field. The proposition is precisely a function or a mark of differentiation. (Prodi 1987a: 54)

Within language, the self interiorizes the others—it *is* the others. Language is something that precedes the self, and that the self can access only when its genetic predisposition to language (i.e. when the phylogenetic history of language has become part of the individual genetic inheritance, predisposing it to an ontogenetic development of language) is actualized by other individual's use of language, starting with one's parents. As an innate predisposition, language is always already this potential space for the encounter with others: upon learning how to speak, the child is immediately in relation with others and therefore becomes another for them, a self facing other selves. Morality begins with knowledge, which is part of the biological make-up of our species; therefore, the individual's entry into language is necessary in order to have moral experiences. Only within language can a complex "federation of readers" become a moral subject, since it is only within language that it can become an individual subject, a singularity capable of posing the moral question upon itself. Morality without an individual would be meaningless, since only the individual can pose the problem of choice and of freedom. Within language, then, the biological potentiality for moral behaviour becomes an effective reality: within language, there is the logical space to form the distinct individuality of a species for which alone the moral problem can be an issue. Language is both norm—and therefore equality—and singularity, the concrete entry of the individual into language, and therefore differentiation:

[w]hile uttering a phrase the individual declares himself as different through it. He is localized by it in his ultimately intentional behaviour. Therefore, the entire communicative-cultural life is conceivable as a continuous flux generating differences — and therefore meanings — that require an impersonal or interpersonal reference to be codified. [...] Therefore, the opposition between codex and individual, freedom and language, or singularity and norm is utterly meaningless. Language is constructed, like a species, via a mechanism of normalization. A norm is also an instrument of singularity. (Prodi 1987a: 157)

Both logical possibility and—in defiance of all those culturalist stereotypes assuming that culture has nothing to do with nature—biological possibility only emerge in language, making possible the appearance of a moral subject: a subject who can choose. It is important to stress this point: Prodi is not defending the dualistic and metaphysical thesis that an ant is determined, while the human being is free. What is at stake is the living being's capacity to perceive itself as a free agent. In fact, an animal is *free* in the moral sense when it can make experience of itself as a free agent. A moral agent is such an agent who can suspend its own actions in order to preserve the autonomy of the object it is aiming to. Such an agent is a by-product of the human animal's capacity to present itself like an "I". The idea is that an *I* can exist only if a *linguistic* "I" exists. This implies again that the complex "federation of readers" that we are—*qua* animal organisms—can reach self-consciousness only through language (as we have started to see in the previous chapter). The Cartesian subject, appearing in the *cogito*, is transparent to itself, an immediate self-intuition preceding anyone else, and indeed a condition of possibility for the existence of others. Coherently with his anti-Cartesian approach (and following Peirce), Prodi sees the *cogito* as a terminal point, rather than as a point of origin, of the process of knowledge:

> the only acceptable formula for the *cogito* would be 'I think *about*, and therefore I am *among*' [penso a, dunque sono tra]. The intentionality of the transcendent I, unifying the object, has nothing to do with the real epistemic interaction where the object is the source of the process. Rather, it is the whole assemblage of objects that constitute the genesis of the subject. (Prodi 1987a: 180)

The subject is engendered by its interaction with others, and not by an intransitive act directly positing the autonomous subject, the Cartesian *cogito*. In particular, the subject is born within language, within a linguistic network that is already intrinsically made of relations and connections:

> by performing this comprehension of external reality, the subject that reads other subjects is actually reading itself. That is, it finds within its own individual existence something in common with the others: the capacity to formulate hypotheses and comparisons, to make choices, as well as to feel joy, pleasure, pain, suffering, and fear. There is no private aspect that cannot be subsumed, albeit partially or with difficulty, to intersubjective relations. (Prodi 1987b: 70)

To be self-conscious means to be able to locate oneself—both as a subject and as an object and as a gaze and as a gazed-upon—within the "theoretical scaffolding" whose boundaries are traced by language. Between these two poles, it becomes then possible to introduce a *hypothesis*, which requires the possibility—once again, logical before factual—of projecting oneself towards a space and a time that are different from those presently inhabited. Only a self-conscious organism can access the space of hypothesis, because in order to imagine (or desire) a different future, it is necessary to be aware of one's existence in a given present. Strictly speaking, without self-consciousness, there cannot be any distinction between times other than the present. In fact, not even the present can be given: a temporal experience flattened on the present, ignoring the possibility of past and future, is non-temporal. There is a present only for those organisms capable of differentiating it from the past and the future. For Prodi, human freedom is intrinsically bound to language and therefore to the quintessentially *linguistic* ability of formulating hypotheses:

> [t]he individual's strategy, within his logical-linguistic domain (and, indirectly, within his objective world) is hypothetical. [...] Every datum coming from the outside goes through the decompression chamber of language — man's internal representational method — and is then translated towards the outside via a hypothetical reformulation. Situations that correspond to possible matters of fact are thus constructed: these are the hypotheses. They are constructed starting with a repertoire, an *a priori* situation — which, in a very general sense, could be called a codex (an intersubjective matter of fact) — by means of our logical -linguistic competence. The hypothesis is first built, and then it is compared with reality. In his

specific actions, man employs this strategy, one that has nothing to do with predetermination. […] The hypothetical strategy means freedom. With freedom I simply mean the capacity of formulating hypotheses. This capacity is constitutive of man. (Prodi 1987a: 47)

The kind of freedom proper to human beings is intrinsically bound to language and in particular to a form of knowledge that amounts to the capacity to transcend things, to look beyond them, to connect them with other things, and to conceive of them beyond their mere use: "freedom is real precisely because it is biological. Human freedom is the capacity of formulating hypotheses" (Prodi 1987b: 54). To know means to read the world via a theoretical infrastructure, a set of hypotheses; and since to formulate a hypothesis means to be inside a language and to know means to use such a language, ultimately to be inside a language and to be—fully and properly—linguistic animals mean being free. This notion of freedom profoundly differs from a more simplistic one which would identify freedom with the absence of limitations to movement and with the simple free choice between a number of alternatives. If these alternatives are not enlightened by consciousness, they are not, properly speaking, choices at all, since any such choice will not be any more meaningful than a completely random decision:

the freedom of those who do not have a mental horizon within which to choose (those who don't possess knowledge and language, or who do not dispose of materials on which propositional knowledge can be applied) is virtually empty. It is, at best, a pure decisional power over other beings: something completely different from freedom as I intend it, to be identified with the physiology of hypotheses. (Prodi 1987a: 47)

For the human animal, freedom is a species-specific characteristic (that is to say, it defines and specifies the human species, *Homo sapiens*), and not a subjective choice; it is given as part of our natural inheritance. The human animal is naturally free because it is an animal with both a well-defined evolutionary past and a peculiar relationship with his language. With a paradoxical conceptual turn, which once again undercuts the traditional opposition between freedom and necessity, necessity is here inscribed in the very possibility of freedom, in the very determinate sense that it is fixed by its biological inheritance:

[the human animal] is determined by freedom. This is not an oxymoron: linguistic competence is innate, but it produces varied and original constructions. The human is forced by its evolution to be, and to remain, an original copy [with respect to the genetic model of its species]. (Prodi 1987a: 52)

Following, once again, the model of the circle, we understand that freedom cannot represent an external supplement to the cognitive endowment of the human animal: as such it would be unjustified, and its origin would remain a mystery. Rather, the human is free because it is determined, because of its peculiar natural and semiotic evolution. Freedom is genuine only within a system of constraints. In order to define freedom, it is necessary to traverse something that is not free, something thoroughly determined like the biological endowment that specifies the human animal as *that kind* of animal and no other. Specifically, if freedom is to be identified with the semiosic capacity to formulate and test hypotheses—and therefore with language— the field of freedom will coincide with that of language: only within language, within the logical space determined by its rules, we can effectively be free.

Finally, the link between knowledge, language, and ethics poses a problem (that I will explore in greater depth in the next chapter) the general contours of which need to be immediately delineated. This problem is a direct consequence of the issue of freedom: what are the limits of our freedom? Our knowledge of the world, Prodi argues, cannot but be filtered through a theory, a network of hypotheses: our knowledge of the world, then, is nothing but another expression of language. Now, according to this characterization, there will always be something that slips through the net, precisely because:

1. A net is so structured to let some things slip through; it is composed by nodes and links, and no matter how close these can be, there will always be holes between them. A net is selective, which means to say that every net will always let something slip through and therefore that any ambition for an absolute knowledge of the world is a pipe dream, a dream in the very precise sense that it is logically (and therefore, for Prodi, from an ontological and semiotic point of view) impossible to hope for an exhaustive knowledge of the world.
2. Every net, as large as it might be, is finite. Too big a net would be useless, since it would not be practical to use. But if every net *has* to be (more or less) small, this means that the sea that slips through it would be, by definition, ungraspable. Knowledge can only cut finite portions of the infinite sea of the knowable. Again, the model of the circle makes this situation intelligible: language, as a generative logical-linguistic apparatus, defines two connceted spaces: the first one is the space of what the language explicitly can formulate; at the same time language implicitly defines an indefinitely extended space of what could be expressed. Since there is no upper limit, at least in line of principle, to the number of admissible linguistic combinations.

But this set, as immense as it can be, is nothing but a fraction of the infinite set within which the rules of our language do *not* apply. The intrinsic logic of language defines two complementary spaces: one regimented by its rules—that coincides with the world, our *Umwelt* (since we have no access to it that not mediated through knowledge/language)—and one which contains the first one, being infinitely more extended, and within which such rules do not apply, as I try to show in Fig. 10.1.

Here we encounter the problem of the limits of the human semiotic field. This can be divided into an aesthetic problem (aesthetics *qua* domain of the unsayable) and a theological problem (intended as the domain of the ultimate meaning of the world). For Prodi, rather than being concerned respectively with beauty and the fear of the unknown, these issues are defined by logic: "the sacred, from this perspective, is a specific drive that is one with logic" (Prodi 1987a: 119). Language, conceived as a moving and expanding circle, defines two concentric spaces: we reside into the space of language, and by definition we are unable to transgress the boundaries imposed by its intrinsic logic, because we cannot think outside of language. Here Prodi is essentially reformulating Wittgenstein's thesis from the *Tractatus* "5.6 The limits of my language mean the limits of my world" (Wittgenstein 1922: 74). Every conceivable access to the world goes through language, and every possible thought is linguistically formulated: it follows that whatever is not bound by the rules of our language is, by definition, unthinkable. "Unthinkable" does not mean that there is

Fig. 10.1 The infinite
space of knowledge/
language and that of the
unknown

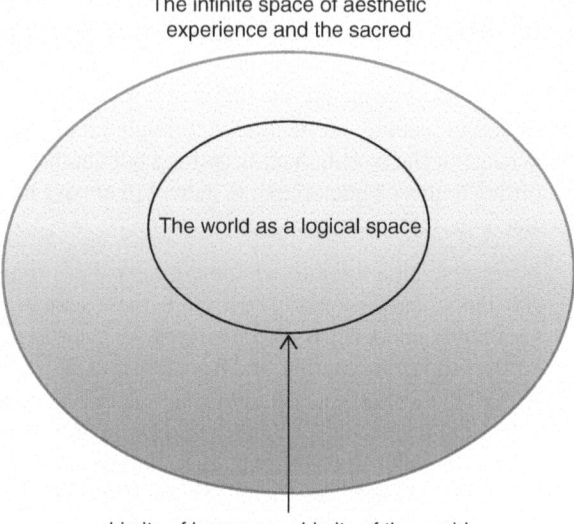

The infinite space of aesthetic
experience and the sacred

The world as a logical space

Limits of language = Limits of the world

another, non-linguistic, way of thinking what the language cannot formulate; it
means that for a human animal is bio-logically precluded the possibility to think
whithout the mediation of language. Since language defines a space, and since we
reside within that space, it bio-logically follows—an entailment that cannot be precisely defined—the existence of something that lies outside the domain of our language. Once again, Prodi reformulates in his own terms the Tractarian problem of
the *internal* limits of language:

> 5.61 Logic fills the world: the limits of the world are also its limits. We cannot therefore say
> in logic: This and this there is in the world, that there is not. For that would apparently
> presuppose that we exclude certain possibilities, and this cannot be the case since otherwise
> logic must get outside the limits of the world: that is, if it could consider these limits from
> the other side also. What we cannot think, that we cannot think: we cannot therefore say
> what we cannot think. (Wittegenstein 1922: 74).

We know, through bio-logical reasoning, that there cannot be any space beyond our
logic, but still this space can be logically said to be unreachable, unknowable, and
unthinkable. It is here that an opening for the logical—but also historical and evolutionary—possibility of those experiences that lie at the limits of language, like the
aesthetic and the religious one, is created:

> [Human reality] is not simply composed by objects like *A, B, C,* and then by some hidden
> objects that are progressively illuminated and captured by language. It is also made up of
> (an ever increasing number of) objects that are not there. Indeed, knowledge is since the
> beginning strictly linked to non-knowledge. If an animal only reacts to the most meaningful
> entities, as if these were the only existing ones, it does not wonder about what might (or
> might not) lie beyond the boundaries of its categories. But when knowledge becomes
> reflexive and propositional, there occurs a contextual emergence of uncertainty about that
> which is not characterized by discourse: the issue of boundaries then arises, *beyond which*
> there lies an unpredictable reality. (Prodi 1987a: 117–118)

Chapter 11
Aesthetic Experience and the Problem of the Sacred

> *The mystery is a matter of fact. An optimistic hope for total clarity is a mirage. Total clarity cannot exist. […]. Therefore, it is appropriate to give the term "mystery" its primordial character, so trivialized by religions and philosophies, that turned it into a mere verbal trick (they know the mystery like the back of their hands, the mystery is for them not mysterious at all). Only the scientific method, and its intransigent character, allows the reconstruction of the original meaning of the mystery.*

(Prodi 1983a: 44)

Abstract The human world is the world of language, i.e. of knowledge. To know means to formulate a hypothesis, to then be either verified or falsified. This also means that the "boundaries" of the world are not fixed but shift with the progress of knowledge. But this also entails that it is knowledge itself that produces the unknown. Prodi calls this movable field of experience, shifting along with the process of knowledge, "darkness" [buio]. The problem of "darkness" is the problem of the internal limits of language. Such "darkness", indeed, is by definition unrecognizable, since it is an inevitable collateral effect of knowledge. There is no science of "darkness", but this does not mean that it cannot be thinkable via other means. Aesthetic experience and the experience of the sacred are non-scientific ways to "think" the "darkness".

Keywords "Darkness" · Hypothesis · Limits of knowledge · Aesthetic experience · Experience of the sacred

In Italy, during Prodi's most philosophically active years, the question of the limits of language was considered a very urgent one. One of Prodi's principal dialogue partners (although never directly cited in Prodi's work) was Emilio Garroni, a philosopher who specialized in aesthetics and Kant (and in particular Kant's *Critique of Judgment*). In his work, Garroni highlighted the transcendental (and therefore, as Prodi would have it, biological) impossibility of achieving an external perspective on the field of language. Figure 10.1, indeed, contains a problem: who can see language from the outside? Whence, to use Prodi's language, the "darkness"? Garroni asks a similar question:

> let us imagine a being who, in order to subsist and to have an experience, needs to be encapsulated — like an insect trapped in amber — in a solid block of transparent material, and that *this and only this* would be its vital and sensorial environment. Well, in these conditions the being that looks is located *within the medium through which* it looks, *and cannot* be outside of it *without* this act of looking to stop being what it is. Let us say, then, that for such a being looking will essentially be a looking-through, at least in the sense that it cannot be possible for it to try and separate its looking and the distortion caused by the medium. This will be impossible to achieve without turning its act of looking into something completely different from the only kind of looking that is, for it, a looking, and therefore without already becoming something *completely different* and, ultimately, *unthinkable*. (Garroni 1992: 12)

In this way, an inexhaustible tension around the experience of the limit is triggered. This is a limit that cannot be logically transgressed but which cannot not to provoke continuous attempts to transgress it: for both Prodi and Garroni, this is the most proper domain of aesthetics. In particular, aesthetics is the field of experience located at the shifting boundary of language/knowledge: being at the boundary, it is at once inside and outside it. It follows that the two philosophers see aesthetics not as a discipline concerned with beauty and art but rather as the experience of the limit, of the "area of the unknown […] a very wide and deep domain" (Prodi 1983a: 11)—this is nothing but the other side of the coin of knowledge. Recalling Fig. 9.1, we should see the domain of aesthetics as the converse as that of science and vice versa.

Semiosis—as we saw in Chap. 4—coincides with life and with biology, and, through an internal process of development and increase in complexity, it becomes knowledge and language. For the human animal, language defines the horizon—the *only* horizon—within which knowable objects, the objects of the world, can be inscribed: a "coincidence between knowledge and system of communication" (Prodi 1982: 200) is therefore instituted. But this system is not closed nor immobile: "every system is evolving, and the evolution depends on the extension of the domain of its use (that is, the enlargement of the contacts with the referent, and the greater complexity of inter-individual discourse, grounded on hypothetical activity" (Prodi 1982: 200). The hypothesis is the proof that the language/world coupling is not stable, since the very need of an hypothesis means that it is not still established the nature of what language is actually speaking. But if it is unstable, it follows that there can always be a new, different way to discover and to think reality, and this is the problem of the limits of knowledge. In Fig. 10.1, both the sayable and the unsayable are represented. The unsayable is such not because something or someone forbids its utterance but because there is (still) nothing that can be said about it.

However, although nothing can be bio-logically said about this indeterminate and indeterminable circumference of the circle of world/language (from Fig. 10.1), this does not mean that it exercises no attraction over us. On the contrary, it is the facet of our experience we truly strive towards, since we hope that we will eventually come to know everything that lies on this side of the boundary: all that we can know through language (the domain of science), i.e. what was always already meant to be knowable. As is his habit, Prodi does not cite anything, but here the reference to the last proposition of the *Tractatus* is clear, where Wittgenstein writes (a claim that with which Prodi could not but agree) that "6.52 We feel that even if all possible scientific questions be answered, the problems of life have still not been touched at all. Of course, there is then no question left, and just this is the answer" (Wittgenstein 1922: 89). To dwell linguistically in the world necessarily means feeling the world as "limited". Only the linguistic animal, the animal that formulates hypotheses, can have this feeling. An ant is always at home in its environment, since its natural make-up makes it impossible for it to ask *where* it is and even less *why*. On the contrary, "with language man is equipped with instruments suitable for dealing with unknown entities" (Prodi 1987b: 57). The hypothesis is the natural instrument used to *feel* the limits of language:

> [r]eality is not simply the sum total of reading machines and of their correlated "horizons of things". These are part of an always wider reality constituted by a — presumably — enormous number of things that are and will be irrelevant for any reading machine, and that will not get the attention of any reader. Within reality, the totality of readers constitutes a network. But they are all still part of a reality that overcomes them, being both wider and more homogeneous than they are. The diversity and peculiarity of various readers cannot be equated with a hierarchy or a preferential relation with reality. The only possible relation towards reality is one of inclusion, with no possibility of inverting the terms. No reader can overcome reality, and contemplate it from the outside. It can only be read from the "inside" of a complex system of interactions. (Prodi 1983a: 17)

Answerable questions are those that are asked from within the circle of language and knowledge; they are not core question precisely because they can be formulated and answered—even when such an answer would remain purely theoretical. Radical questions, those that involve our entire being, are those that cannot be asked due to the lack of the necessary linguistic tools: they are located beyond the rules according to which we—and we alone among all animals—can articulate them. For this reason, within that space, "there is then no question left" as Wittgenstein wrote, since it is impossible to formulate bio-logically a question regarding a region located beyond human bio-logic. This means that there is only one thing that can be said regarding the questions that cannot be asked: *that they cannot be asked*— "and just this is the answer". It is important to specify that "they cannot be asked" does not mean that there is someone who prevents us to pose these questions; according to Wittgenstein the very bio-logical nature of human beings prevents us to pose these question. *Homo sapiens* is such an impossibility. However, according to Prodi this solution is unsatisfactory, since it appears to be the mere acknowledgement of a limit and of an impossibility, and not a positive solution for the unease that prompted the questioning.

Within the circle, we are attracted by what lies beyond its borders, precisely what, by definition, we cannot know, since the logic intrinsic to our life and our language forecloses the knowledge of that which we cannot understand, that lies beyond the borders of language, and that escapes the grasp of our categories. So, Prodi writes that:

> the fundamental problem is that of borders: what we cannot understand, that lies beyond the limits, and for which we have no categories. In our epistemology and our ontology is

implicit that the borders of the world are not the same as the limits of our ways of knowing it, and that if silence is an appropriate response regarding that which we cannot say,[1] it is however necessary to acknowledge this silence, i.e. to believe that there can be something that, for us, will forever remain silent. (Prodi 1987b: 212)

As we have seen, this indefinite region of extralogical space can only be inferred from within human bio-logical space:

reading machines, according to their degree of complexity, can read a different spectrum of reality and even — as in the case of man — move very far from the domain they originally belonged to, but always traversing the modes and the links of the internal relations of the real. (Prodi 1983a: 17)

This area beyond language, which is still somehow "felt" without ever leaving the circle of language, can be approached through two distinct yet connected methods: via a hypothetical process of extension of language itself, trying to reduce the unknown to the known, or by wholeheartedly accepting the logically inevitable fact of the limits of language/world. The first of these two methods is the aesthetic perspective—in particular, the one offered by poetry—while the second is that of the sacred.

Let us go back to the circle of language/world, as represented in Fig. 10.1. The human animal, in virtue of its being human, in fact coincides with its language. And language draws the boundaries of its world since—following the biological complementarity between reader and thing read—the world is that which is experienceable/knowable: "an organism knows/interprets (has a specific relationship with) the reality *upon which* it has constructed itself. It interprets the world through its categories; but these categories have been constructed by *the world itself*" (Prodi 1987b: 143–144). And since the set of all knowable things coincides with that of language—"there is no qualitative distinction between everyday and scientific discourses" (Prodi 1987b: 102)—the limits of the world coincide with the (actual) limits established by the rules governing what we can think and say. The point is that, as we have already seen, our human experience is wholly internal to that circle; yet this does not stop us from feeling that beyond these borders, where our categories fail, there is an outside—albeit unknowable —that attracts us, makes us uneasy, and somehow compels us to explore and to know it. This feeling originates, once again, within the circle, precisely because we know that our knowledge—i.e. the sum total of the operations we perform in order to recognize the meaningfulness of some portions of the real (only *some* portions, since the whole of reality is by definition ungraspable by beings that can only inhabit a single point of view, the human *Umwelt*)—acquires a meaning only against the backdrop of the larger and infinitely wide region of the unknown. Prodi observes that this is:

a fundamental element of discourse, and holds the weight of all those things that exist but about which we have no proof. So, the word implies the non-word, not by means of dialectical games but through the mechanism of the genesis of discourse. (Prodi 1983a: 40)

---

[1] The reference is to the last proposition of the *Tractatus*: "7 Whereof one cannot speak, thereof one must be silent" (Wittgenstein 1922: 90).

Knowledge is the natural operation through which the meaningfulness of something is established—and this entails the exclusion of a multitude of other things. That which is not meaningful remains in the background, just like every word defines itself against a much wider space of silence, unknown and unthought. But this entails that the unknown is somehow always present, as a background or infinite horizon of (possible) knowledge:

> if we think of "rationality" as all those ways in which man learnt how to "humanly" handle a certain portion of reality (thus making it a little more his own), then the rational is only a tiny portion, cut against the background of the irrational and always placed within it. The rational is a minuscule "organized irrational". (Prodi 1983a: 41)

Aesthetic experience, which Prodi does not consider limited to that of beauty or art, is the kind of experience located in the vicinity of the border. It is akin to knowledge, but it represents its most initial, hypothetical, and tentative moment, precisely because it is originated by a "drive for knowledge".[2] An aesthetic experience occurs when, starting with the experience of a precise cognitive placement within the circle of language/knowledge, one tries to move beyond this place and to proceed towards the outside, the unknown, or—as Prodi defines it—towards the "darkness":

> the mystery lies [...] deep inside. If the world is far wider than our conceptual frame and dwarfs us both in terms of its physical dimensions and its history, then our being included within it is the crucial node of existence and of consciousness. There is always an outside, an area of darkness, and it is vain to think that this will ever be exhausted. (Prodi 1974: 169)

As we will see later on, the "darkness" is nothing but another way to allude to that which the sacred refers to with the term "God", something unknowable and always exceeding our limited descriptive resources:

> [t]he hard core of the aesthetic feeling is a certain undecidable placement, our being integrally the object of this undecidability that characterizes the limits (and their beyond). We presumably are already beyond these limits, yet we are also radically opaque to ourselves, unable to perform a thorough introspection or to take shortcuts — while being still in proximity of ourselves, almost coinciding with our own centre, if there is one. (Prodi 1987b: 216)

Aesthetics, as a primordial and permanent form of consciousness and experience of our being limited—logically limited yet (bio)logically driven to exceed the limits—coincides with an area of doubt, of questions without an answer (yet). Put differently, using Wittgenstein's terms, it is the area of the *awe* felt when facing the world *qua* world: "[a]esthetically, the miracle is that the world exists. That there is what there is" (Wittgenstein 1998: 86). This awe always accompanies the feeling of being trapped within language, like Wittgenstein himself observed, in his *Lecture on Ethics*:

> [f]or all I wanted to do with them was just *to go beyond* the world and that is to say beyond significant language. My whole tendency and I believe the tendency of all men who ever

---

[2] It is apparent that Prodi is referring to the "Wissen Trieb" in the *Three Essays on the Theory of Sexuality* (Freud 2000). According to Prodi and Garroni, aesthetics has more in common with science than it does with art.

tried to write or talk Ethics or Religion was to run against the boundaries of language. This running against the walls of our cage is perfectly, absolutely hopeless. Ethics so far as it springs from the desire to say something about the ultimate meaning of life, the absolute good, the absolute valuable, can be no science. What it says does not add to our knowledge in any sense. But it is a document of a tendency in the human mind which I personally cannot help respecting deeply and I would not for my life ridicule it. (Wittgenstein 1993: 44)

The human animal responds to the unease produced by the experience of the limit with a natural skill designed to face uncertainty: the formulation of hypotheses, the attempt to extrapolate to the unknown and to the future what is already known about the present and the past. One can formulate this point in a different way: the basic bio-logical fact that we mainly approached the world through the mediation of language means that human beings do not live into a *Umwelt*. A *Umwelt* is a *Umwelt* just because it does not pose any question to the animal who lives inside it. On the contrary, the human world presents itself as something which is not at all immediately intelligible; therefore we need do find a meaning to what we perceive into 'our' world. The existence of language is the proof that we do not understand the world we live in. Prodi indeed writes that:

hypothesis and doubt are two aspects of the same mechanism. [...] It is the hypothesis [...] that transforms the knowable world. Doubt refers to areas of the real that are much wider than those we have experience of: it pertains to that domain of "meaningless" things that could become meaningful, an area of possibility. This is not a simple psychological state. It is a *res extensa* that can be linguistically organized. (Prodi 1983a: 38)

In truth, the hypothesis is an internal instrument of language, making our experience of time possible: it decentres the speaker with respect to the moment of his or her utterance, dislocating him or her to another time—the time of hypothesis. Thus, the hypothesis opens the possibility of moving through time, since it breaks the identity between the moment of enunciation and the present time.

A hypothesis is not (and perhaps it never can be) justified, precisely because it is a hypothesis, i.e. an attempt to extend the boundaries of the known into the domain of the unknown. The hypothesis is such precisely because it is unjustified: if it was justified, it would not be a hypothesis, and it would instead be a proposition expressing one of the many accepted and scientifically described facts that lie within the circle of the world/language. So, the hypothesis is a hybrid construct, suspended between the past and the future, between ascertained knowledge and possible future knowledge, and between memory and anticipation: hypotheses are "internal linguistic organizations of memorized data" (Prodi 1983a: 210). In fact, a hypothesis is nothing but the possibility to experience time, because its own "time" is the future; therefore, the "future" shows itself into human "mental" life through the hypotheses. But when the "future" shows itself, the "present" and "past" times also show themselves, because a "future" can only exist in contrast with other times. This is another by-product of the fact that language in fact concides with the human world: we can formulate hypotheses, thanks to the combinatorial resources of language, allowing us to internally generate propositions not (yet) dependent on the characteristics of the external world. A hypothesis is (relatively) free, since language, as a combinatorial machine, allows the production of an infinite number of utterances, each of which can represent a possible exploratory strategy, a way to grapple to contents—beyond the space of world/language—that were simply waiting to become (in Prodi's terms) meaningful, i.e. to select a reading machine able to read them:

> [the hypothesis] is an original nucleus of imaginative elaboration [...]. The hypothesis is a fantasy scenario, by which I mean a brain state characterized by the subject's capacity to freely employ linguistic material he owns, without depending on any selection, restrictions, external/environmental censorships, or feedback onto the real. The hypothesis is a material state of the structure. The internal space extends into the external space, the pole of consciousness: but during the hypothesis' operative phase the former is temporarily isolated from the latter. (Prodi 1983a: 209)

A hypothesis is intrinsically epistemic but also intrinsically aesthetic (poetic) since it is an internal experimentation, a play of language that appears to function as if moved by a purely combinatorial logic. Yet, it produces what we could define the "cognitive tentacles" that tear fragments of the unknown away from the "darkness", in order to make them sayable/knowable. By means of a hypothesis, we build a fantastic bridge resting on only one shore—that internal to the world/language—and suspended on the other or perhaps (paradoxically, for this is the intrinsic paradox of hypotheses) resting upon itself. Hence, Prodi writes:

> the path of discourse does not terminate with a verifying procedure (a final demonstration), but it is always seeking further verification, i.e. a form of knowledge that would be established through its progress, and through its character of experimental testing, of provisional attempt. (Prodi 1983a: 216)

This paradoxical condition derives from the logical impossibility of speaking about that which lies beyond the borders of the world/language. By definition, nothing meaningful can be said about what is beyond these limits, precisely because no evidence can be brought to bear upon the truth or falsity of our propositions about it. Yet we can still talk about it, and, historically, a lot has indeed been said about it, by means of hypotheses, conjectures, and analogies:

> it is certain that *nothing* can be said about the unsayable. [...] However, *if not about the unsayable qua referent/object, we can talk about the unsayable qua sum total of our experiences about it,* real human experiences, since man has always talked about the unsayable. (Prodi 1983a: 42)

Science, coextensive with language—the two being both evolutionary developments of simpler forms of semiosis and therefore of life—defines the boundaries of our world. That is to say, only within science's—and language's—sphere of action can we identify known and well-defined objects. However, referencing Wittgenstein, we have also said that science does not exhaust our experience of language nor does it placate our unease towards the domain of the unknown, the "darkness". Our eyes can only look inwards, so to speak, but at the same time, we feel a very strong "epistemic compulsion" towards that which lies outside, that we cannot see, and yet know that—bio-logically speaking—*cannot but exist.* There is an inside—something which has been already said—only because there is an outside, something (still) unsaid, something extralogical. The only (bio-logical) method we have to access this domain is through the formulation of hypotheses, through the use of analogies, and in general through the aesthetic use of language:

[a]ll we have about these domains is the possibility of speaking (and having spoken) about them: that is to say, the linguistic experiences we have accumulated. [...] When the problem is radicalized, and we tackle the issue of the representation of our worldly placement via an explicitly linguistic perspective, we can see that this allows a dynamic representation: it is experimental, original, communicable, progressive, and it delivers knowledge about ourselves. It is poetry, in the broadest possible sense. *The only alternative to scientific language, in the domains where the latter cannot be applied, is another linguistic structure (or better, another use of the linguistic structure) that experiments not with things, but with segments of language made meaningful through their use.* This is poetry, language employed aesthetically. One could say: "but then it is impossible to say anything about the unknown as a reality, unless this can be penetrated by the scientific method". That is exactly right. [...] But it is also possible to utter words about the unknown, and about our way of living it. (Prodi 1983a: 43–44)

Let us consider Fig. 10.1 once again. The aesthetic dimension of language, i.e. the hypothetical use of language, is located at the boundaries of our world. Better still, it stands on the boundary, a place where we logically could not stand but where we try to reside by using the instrument of analogy. Analogy is that peculiar linguistic procedure through which what is known is projected onto the unknown in order to try to test its effectiveness and to see if that portion of "darkness" considered by our hypothesis is regulated by the same laws that apply in the domain of the already known. The hypothesis, then, functions analogically:

the analogy operates aesthetically towards the darkness, as an attempt of translation/enlargement of our terrain towards it, in terms of verbal constructs. These cannot be verified; they have no meaning in their objective referentiality. They are grounded on a radical analogical mechanism: [...] to lend our point of view to the darkness, to be seen by unseen eyes. To be seen by things rather than see them, even from their hidden side. (Prodi 1983a: 207–208)

This means that, thanks to the aesthetic use of language, the boundaries of the world/language can be extended: at first, simply hypothetically and then—if the analogy works—also cognitively, as I have tried to represent schematically in Fig. 11.1, where the analogies/hypotheses project towards the unknown (the "darkness") some kind of "heuristic tentacles". The domain of the "darkness" coincides with that of aesthetics and the sacred and therefore also with art and religion, art as the material instantiation of the aesthetic function and religion as a reassuring "answer" to the *logical* problem of the "mystery":

[e]ssentially, it is meaningless to speak of the unsayable *qua* object: but we can still say its unsayability. Just as I have called the religious problem, strictly speaking, the problem of the boundaries (our only way to live an object about which nothing can be said), so I will talk of aesthetics or poetry when referring to what pertains to our (linguistic and epistemic) human experience of the problem itself. Since the beginning, man has projected his experience upon the unsayable, the mysterious, that which is seen as laying beyond the borders: this has then become linguistic material. It was transformed — and continuously transforms itself — into those semantic/syntactic constructs that compose the linguistic reservoir from which we take the materials for our discursive hypotheses. This is a non-trivial part of language. Indeed, it is its most crucial part. [...] The sacred has shaped man no less than hunger. (Prodi 1987b: 216)

The analogical procedure, which characterizes the specifically aesthetic mode of approach to the real, is crystallized in an act of nomination through which a new

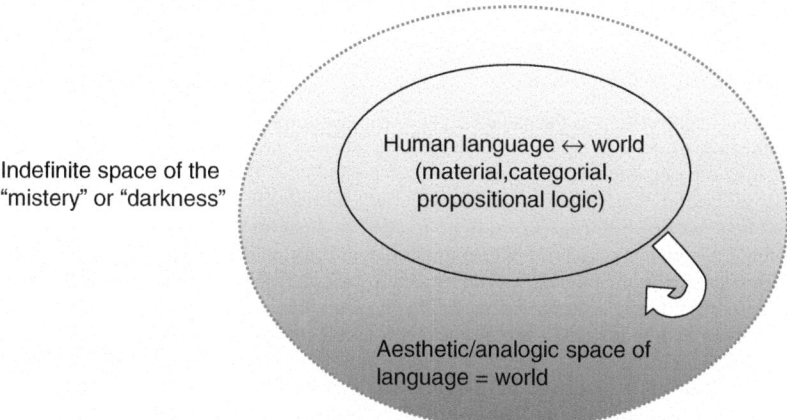

Fig. 11.1 The aesthetic function as hypothesis that ventures into the "darkness"

entity is introduced in the world/language. In this way, an entity is subtracted from the domain of the "darkness" and inserted—albeit in a hypothetical and temporary manner—within the circle of language, of knowledge, and of science. The fundamental link that allows the word to become a kind of ontological *passe-partout* is established between the word and a voluntary action. The nominated object is not necessarily a concrete one, and indeed it likely is—and here we witness the progress of knowledge—an immaterial theoretical entity. However, it becomes the object of our attention the moment that a name tears it away from the darkness and places it in our epistemic spotlight, making it both visible and knowable. The crucial point is that such an entity does not properly exist as a thing *before* being aesthetically (hypothetically) summoned. That is, an object receives a name and receives the status of "object" only through the aesthetic function of language. This does not mean that such an aesthetic function "creates" an object; it means that it becomes something worth paying attention to through the mediation of language. What is at stake here is not the material existence of the object; the point is how to push it, whatever it is, into the human world. It is appropriate to specify that the "darkness" only exists for the human beings, those animals who make experience of their world through language/hypothesis.

The aesthetic function anticipates the existence of a nominated thing, because objectification *follows* nomination, not the other way around. As we have already seen, this does not mean that the name *creates* the object, since we know that every name is nothing but a narrow cut-out extracted from the enormous domain of the unknown. The nominated object or event presumably existed before its nomination, but it was unknown in the darkness (that is, the darkness is nothing but such a biological unknowableness), and therefore its existence was—for science—unknown and at best object of a conjecture. It was there as a potentiality for knowledge, but it was not yet meaningful for us; it was not among the objects of our world. Only through analogy, and the act of nomination that represents the (momentary) conclusion of the analogical process, can such an object or event reach its ontological fullness and be admitted within the circle of language/world. The name, strictly speaking, does not create the thing, but neither does it merely register its existence by assigning it an utterable label:

> [t]he misinterpretation of the name [...] as a reproduction of the thing, and of discourse as a miniature of the real, derives from the consideration of the act of nomination of a reality as superimposed on to the thing, as a pure correspondence. There is an existing thing, and *then* we label it with a name. [...] On the contrary, the name is not reducible to a repetition, a static replacement for the thing, to be inserted in a conventional type of relations. The name is the product of a series of operations that have, as their focal point, the object or the phenomenon. It is their operational equivalent. It "stands for" in a completely different way than a conventional construction that would once and for all establish an equivalence. [...] The name recapitulates an inquiry into reality. With a name we bring to a conclusion a series of operations of verification: it is therefore an experimental system of translation. [...]. The name is the equivalent of the operations that had the thing or phenomenon as their primary concern, and that have triggered and "reduced to one" our discriminatory functions with regard to the real. (Prodi 1983a: 186–187)

The "darkness" is for us both a source of unease and of attraction; we feel that in the "darkness", there are an infinite number of new objects—and therefore new kinds of knowledge, mostly about ourselves—because we are nothing but one of the many shapes taken by life, which is a form of the "darkness" itself. For this reason—the principle of continuity that links together all forms of life and semiosis—to know the darkness means to know ourselves. But before knowing them through scientific methods, we need to somehow present these "dark" entities to our attention (and, for those farther still from the borders of our language/world, scientific exploration is so far ahead in time to be irrelevant to us, and in these cases aesthetic appropriation will suffice). This is the purpose of analogy, of the name in which the aesthetic-analogical movement finds a temporary resting place: to focus our attention on that entity. Or better still—since strictly (cognitively) speaking that entity still does not exist—it serves to unify, or to give a target to, the whole of our activities, blindly addressed to that entity (and prompted more by the *hope* of finding it than by the certainty of its existence). Prodi writes that:

> it is the process of convergence towards an object that has the dominant effect on the act of nomination, since it focuses our eyes and our intelligence towards an external centre. This process terminates in a name, which is not the equivalent of the thing, but of the series of operations that, once triggered by the thing, proceed from us in order to individuate it. (Prodi 1983a: 188)

If analogy—poetry and art as a whole—is a form of knowledge, then the problem of its epistemic value arises: are the constructs of poetry somehow true or verifiable, or are they nothing but useless and meaningless wordplay? In order to answer this question, it is necessary to retrace the genesis of analogy. We know it to be, essentially, a hypothesis: this is one of the forms through which language manifests itself. Indeed the hypothesis is intrinsically linguistic (since for an animal who does not think through language, Prodi argues, it is bio-logically impossible to formulate hypotheses). Human language represents the complex evolutionary outcome of a vast number of different forms of semiosis, pervading the entire biological world which—properly speaking—indeed coincides with semiosis. Therefore, the linguistic nature of analogy is—both logically and ontologically—derivative from the intrinsically semiotic nature of the world. And for this reason, analogy, although nothing more than a free linguistic combination and wordplay, can grasp the real and therefore be true—or better, it can function as a hypothesis for the construction of genuinely true theories. Now, according to the reciprocity entailed by the circular model, an analogy can also be seen as starting from the other side of the circle: that is to say, not just from the subject towards the "darkness" but also from the "darkness" towards the subject. In this case, analogy could be considered as the "darkness" trying to become transparent to itself, according to that already-mentioned, and rather unusual (to Cartesian ears), definition of the human offered by Prodi: "nature thinking itself, the interiority of nature" [natura che si pensa, l'interiorità della natura] (Prodi 1987b: 93).[3] The analogy is real because it is a manifestation of the real itself: it is a hypothesis of translation linked, via an indefinite series of "translations chains", to the real itself that, as a whole, coincides with the infinite network composed by these chains:

> [t]he existence of this kind of translation has a natural reason, even though, considering the absolute lack of a response coming from the domain of darkness, we only seem to hear a faint echo of our own questions. Whatever the darkness might be, we understand it from within, since it generated us, and we are included in its reality as particular cases. We are simply a small region of darkness that became systematic, linguistic, and — perhaps minimally — self-conscious. There exists an aesthetic relation with the darkness because we have an objective placement within it and a genetic relation with it. (Prodi 1983a: 208)

If aesthetics is akin to science—since in both cases it is necessary to "digest the unknown through the digestive system of a functioning language" (Prodi 1983a: 208–209)—there remains the logical (and therefore unavoidable) fact that neither experience can, as Wittgenstein reminded us, ease our anxiety when faced with the indeterminate expanse of the "darkness". Indeed knowledge, whether scientific or aesthetic, emerges precisely from the attempt practically to overcome this unease. Yet, due to its partial and always incomplete nature, this knowledge does nothing but increase our dissatisfaction and our feeling of finitude. The endless immensity of the unknown always extends in front of our *logical* eyes, as well as our physical eyes, with which we contemplate the infinity of the starry sky at night and which makes us feel limited, nothing but an infinitesimal part of an infinite "darkness". This is the "extended [...] region of the unsayable [...] overwhelming us from the outside" (Prodi 1983a: 195), that many—employing a word laden with an anthropomorphism wholly alien to Prodi—refer to as "God".

For Prodi, this word does not refer to a particular substance nor to a person: if anything, it is the possible matrix of *every* determinate thing, and in particular it represents the overarching, mute horizon of our dissatisfaction as linguistic animals. Linguistic animals who can produce hypotheses because, through language, we can reach the awareness of living inside a closed world, albeit one infinitely extended on the inside: "6.432 *How* the world is, is completely indifferent for what is higher. God

---

[3] This formulation seems to suggest that Prodi would not be against the hypothesis of an "anthropic principle" (Barrow and Tipler 1988).

does not reveal himself in the world" (Wittgenstein 1922: 89). For Prodi, that of the the sacred is a bio-logical, and not a cognitive or emotional, problem (that is, theology cannot exist without biology). It is a logical problem because the functioning of our language/hypothesis repeatedly reminds us of the limited nature of the world. In this sense, the "darkness" is a by-product of every linguistic act, even the most banal and unremarkable ones: hence, the "darkness" is not a psychological—private and subjective—issue. Rather it represents the objective correlate of ordinary language (once again, think of a Möbius strip). The "darkness" emerges from an ineliminable *logical feeling*—and this oxymoron effectively encapsulates the originality of Prodi's stance—that we are located inside a circle. This is the circle of the world/language that cannot be observed from the outside, since this would entail the abandonment of our logic and our language and the adoption of new ones. But naturally (logically), this is impossible, since we are sons and daughters of this language and cannot abandon it at will: which is to say that language (and the human animal, deriving from it) cannot define itself. From this also follows that we are *bio-logically* finite beings, in relation to an infinity that exceeds us, overwhelms us, and towers above us. The problem of the sacred, then, pertains more to bio-logic or bio-semiotics than to the sphere of feelings, since it is through the former that we become conscious, both practically and existentially, of our limits: "the sacred, from this perspective, is therefore a precise drive that is one with logic" (Prodi 1987a: 119). In this sense, the problem of the sacred (unlike the *religious* problem, pertaining to human subjective beliefs belonging to determinate historical religions)—*pace* those utopic materialisms presenting a world without "God"—is inscribed into our very being:

> [a]daptation means organized and unitary reading of the environment. This unitary and organized character is always linked to the peculiar and punctual nature of the organism. There never is a [...] *generic* reading, a reading of the *entirety* of reality. Therefore, the organism never emerges out of (i.e. does not correspond to) reality as a whole, it does not stand in front of reality in a biunivocal manner (as large as reality, or as engendering reality, or as capable or interpreting the whole of reality), nor does it present itself as *opposite* to reality. Its position is one of inclusion, dependency, and embeddedness. It is genetically conditioned and adapted (as derivation and knowledge) to a part of reality. (Prodi 1983a: 21–22)

Materialist and functionalist theories of religion try to explain its origin by connecting it to a feeling of dread, the fear of the unknown, and therefore relegating it to our psyche's most obscure and blind drives. Religion would then be nothing but a sentiment that—at least in principle—could one day be made obsolete by education and sufficiently developed scientific knowledge. On the contrary, following Prodi's approach to the problem of the sacred, the latter is to be seen as inscribed in our biological make-up, not in the banal sense that our idea of "God" would be somehow innate but rather meaning that our *radically linguistic* nature brings us, through an aesthetic and hypothetical experience, to interrogate ourselves about our limits and to fantasize about that which can or cannot lie beyond them. Once again then, the experience of the sacred is akin to an instinct, an extremely peculiar *logical instinct* (another oxymoron) which is an integral part of our being intrinsically and radically linguistic animals:

[a]s a whole, the hypothesis I propose goes against anthropological and ethnographical theories grounded on the absolutely *irrational* originality and primordiality of the sacred, elements that would be proper of an under-developed phase of a civilization. These invariably start by considering a man as already genetically constituted. Here, instead, I want to argue that the origin of the sacred needs to be backdated to the phase of the rational constitution of man, to the creation of his specific differential traits. It is contextual to logic and to morality. The sacred characterizes the passage from the non-man to man, and contributes to the creation of his specificity. (Prodi 1987a: 118)

Of course, Prodi's "God" is not the personal deity of Western theological reflection, and it is a void more than a being, an abyss. Correlatively, rather than outside of us—as it bio-logically is—this "God" is located inside us, generated as a biological consequence of the way in which our world as multilayered linguistic animals is, both epistemically and linguistically, constituted. It is within us in the sense that a religious problem cannot possibly emerge within any other animal than us, since non-human animals lack the cognitive/semiotic resources necessary to project themselves through hypothesis/analogy beyond their immediate present and beyond the limits of their environment. Without this capacity—which presupposes the painful and exalted experience of the limit—it is not possible to grasp one's finitude fully. To put it in another way: the problem of the sacred presents itself to us just because our world is not properly an *Umwelt*. In fact an *Umwelt* is closed, while human world, as a consequence of language/hypotesis, is indefinetely open. A non-human animal cannot experience the bio-semiotic feeling of being trapped into a closed space: although it lives—objectively, i.e. to our eyes—in a world and as part of a given environment, it is not aware of it. The non-human animal is intrinsically a-religious (not atheist, nor a believer), since in its life form, there is no logical-linguistic device that makes the emergence of the problem of the sacred possible. For such an animal, "God" as instantiation of the "darkness" cannot possibly exist: "darkness" only exists as a condition of absence of light, but not as a question, as an interpellation. However, one could say that just because animals do not explicitly believe in "God" they are the only *true* believers. They could be described as so completely absorbed by God that they do not need to believe in "God". Anyway, in the end we return to the circle and the continuity that imperceptibly links together all the points of its circumference: the problem of the sacred is, at bottom, an intrinsically biological one, deriving from life and perceivable only within its world (and this, it should be clear enough by now, is not reductionism, a naïve naturalism, or any stance simply looking to identify the foundation of our behaviours in our animal condition):

[t]hus man learns that, since doors can be opened, other unsayable rooms and corridors extend beyond them. And it is precisely from knowledge, as a propensity to the sacred, that the contextual attitude towards the "mystery" that belongs to it emerges. The darkness as a mute presence, not revealing anything but standing beyond everything and whose precise placement is unknown. We have time and again denied that scientific knowledge could be conceived as a light banishing *all* shadows with its splendour. Rather, it is part of the meaning of the mystery, proper of a species that can reflect upon itself, and that has constructed itself thanks to such a self-reflection. (Prodi 1987a: 120)

Chapter 12
Conclusion: Prodi and Italian Thought

> *There aren't two kinds of things (natural things and human things) to be observed from two different perspectives, but only one kind. What differs is only the degree of "mediation" needed to reach them. In the case of human things, the path is much longer. I mean to say that the human sciences are part of the science of nature; conversely, this also means that the natural sciences are far more human than is normally believed.*

(Prodi 1987a: 175)

Abstract In this final chapter, I will place the figure of Prodi as a natural philosopher in the wider context of Italian philosophy. Indeed, his originality and his commitment to science notwithstanding, Prodi has always been a philosopher engaged, in a more or less explicit manner, in dialogue with the Italian philosophical tradition, which in recent years has defined "Italian thought". It is impossible to understand Prodi and his timeliness unless his work is read in the context of this tradition, whose primary characteristics are, on the one hand, a radical anti-dualism and, on the other, naturalism, yet one not to be confused with bare materialism.

Keywords Transcendence · Immanence · Italian thought · Naturalism

In his books Prodi almost never cites other philosophers, and when he does, it is often in order to criticize them. His books are long, and at times somewhat convoluted, sequences of thoughts that seem not to have precedents nor to leave space for successors (and this is not the only issue that has limited the dissemination of his work). Prodi appears as an isolated figure, without a clear provenance nor direction. It is perhaps for this reason that his thought has been so little explored posthumously, as though he was not an original thinker, and his apparently stubborn attempt to reconcile science and philosophy, biology and culture, genetics and language was both misguided and out of step. That is, maybe Prodi is interesting, but uncategorizable, and always too far from mainstream debates. At a time when everyone was involved in the "linguistic turn", he was talking about chemistry and

© Springer Nature Switzerland AG 2018
F. Cimatti, *A Biosemiotic Ontology*, Biosemiotics 18,
https://doi.org/10.1007/978-3-319-97903-8_12

biology; when, shortly before his death, the topics of sociobiology and the "natural-ization of the mental" were beginning to gain traction, he was still defending the specificity of culture and of human language: always out of fashion, always untimely. However, in this chapter, I will try to place Giorgio Prodi in his temporal and cul-tural context, proceeding in a double direction. In the first, I will examine the par-ticular Italian cultural climate within which his scientific and philosophical activity takes place (i.e. in the second half of the twentieth century); in the second, I will consider a tradition within a longer timeframe, extending back in time all the way to Dante Alighieri, one to which Prodi belongs as an Italian intellectual.

Let us begin with the one closest in time. Which cultural environment was addressed by Prodi's 1977 book *Le basi materiali della significazione*? This was Prodi's most innovative monograph, at least with respect to the Italian philosophical context of the period. In an interview in 1986, Prodi describes the reception of his book thus "as a whole, the Italian environment to which my book was explicitly addressed, was thoroughly hostile to the kinds of problems I wanted to tackle. It was rather obvious that the book would not find a receptive readership within the philo-sophical landscape composed by Crocians, former Crocians, Marxists, former Marxists, and irrationalists of various stripes and provenance. In Italy, philosophers attentive to problems from the history of science have a nearly-exclusive interest in archival issues" (Prodi 1986: 123). This is a stern—and perhaps somewhat ungener-ous[1]—diagnosis of his time but on the whole correct. The twin influences of Marxist and Crocian historicism on the one hand and of Catholicism on the other did create obstacles for a productive encounter with the new scientific discoveries of the 1900s. In this framework, semiotics represents a peculiar case. This was a discipline that—through Umberto Eco's mediation—paid some attention to Prodi's work (we should recall that almost all of his major monographs were published by Bompiani, a pub-lishing house for which Eco used to be one of the principal editorial advisors).

In order to understand fully the cultural climate of those years with particular reference to semiotics, I will discuss two concepts from Eco's *Trattato di semiotica generale* (1975, translated to English in 1976 as *A Theory of Semiotics*) that Prodi could not but find problematic[2]: that of "referential fallacy" and that of "unlimited semiosis". Let us analyse the first of the two by referring to Eco's famous example, the case of a codex giving information on the water level of a well. What is at stake is if the *actual* presence of water is relevant from a semiotic point of view. In such a case, when speaking of "water", the existence of water, according to Eco, "even though it certainly was a necessary condition for the *design* of the model, it is not a necessary condition for its semiotic *functioning*" (Eco 1976: 58). According to Eco

---

[1] For example, it has now become fashionable to blame Croce and Gentile for the backwardness and provincial nature of Italian culture (particularly that of the sciences). In truth, this is a com-monplace but inaccurate caricature, especially so in Croce's case, one of the few Italian intellectu-als who truly broke out of the inwardness of Italian culture in the early twentieth century, particularly during the dark fascist decades.

[2] However, one should here praise Umberto Eco who encouraged the publication of Prodi's works, even while these often contained theses that were opposed to his own.

we are here in the presence of another threshold, that "between *conditions of significations* and *conditions of truth*, in other words the threshold between an *intensional* and an *extensional* semantics" (Eco 1976: 59). The second concept is that, even more controversially, of "unlimited semiosis". Eco explains that:

> a cultural entity never obliges one to replace it by means of something which is not a semiotic entity, and never asks to be explained by means of something which is not a semiotic entity [...]. *Semiosis explains itself by itself*; this continual circularity is the normal condition of signification and even allows communication to use signs in order to mention things. (Eco 1976: 71)

Umberto Eco was certainly not a Marxist historicist nor a Catholic thinker. However, it is clear that, if thus conceived, semiotics becomes a self-sufficient sphere of human activity, wholly separate from its material preconditions. In this sense, it is possible to speak, regarding Italian philosophical culture in the 1970s and 1980s, of a certain more or less explicit idealist tendency. This kind of semiotics has completely detached itself from its biological origins and the rest of the living world. On the contrary, as I have stressed in several occasions, Prodi is a philosopher of continuity, and he never accepted this radical separation of the human from the rest of the living world:

> [T]his is how I see the problem: the relation between philosophy (the old monopoly of theory) and science (the concrete emergence of the empirical datum and the scientific method) is today critical. But in order to resolve this issue, in the final analysis, the proposed solutions are always the same two: 1) philosophy is dead or 2) it is necessary to improve the relation between the two domains. These are clearly unsatisfactory solutions, as they do not say anything at all. The first is a funerary elegy for philosophers (for philosophical researchers it matters very little if the philosophy they work on is living or dead). The second produces a large number of cultural entertainers, ready to organize social happenings and to take ready-made formulas out of their hats: from admonitions like "if only humanists knew the second law of thermodynamics and scientific researchers knew Petrarch!" to the invention of new hybrid vocabularies, transforming energy in spirit, and the uncertainty principle in free will. These are neither solutions, nor mediations (they cannot even support the myth of interdisciplinarity). [...] Today, true contact between the two domains (within the concrete perspective forced upon us by science) can be had only where specific experimental data can be employed for the resolution of equally specific traditional philosophical problems. (Prodi 1986: 124)

Prodi's originality should again be underlined: he is proposing a real collaboration between science and philosophy as *equal* partners. From this point of view, his proposal should not be confused with the now-fashionable "experimental philosophy" (Knobe and Nichols 2008). The philosophers should not turn into a lesser version, unconstrained by the mathematical formalism and experimental work of the scientist. Rather, it is necessary to develop a wholly new field of research, what Prodi calls "general semiotics" (and today known as biosemiotics). In this case, Prodi writes, "it is not a matter of interdisciplinarity — a failed project whenever it has been attempted — nor of polite relations between neighbouring corporations: what is at stake is a new language, a unitary perspective, not comparable with any of the two views it seeks to merge. That is to say a wholly new way of configuring the problem of knowledge" (Prodi 1986: 125).

In order to account fully for the lack of attention given to his philosophical-scientific work, one last aspect of Prodi's intellectual profile needs to be highlighted: he always rejected any totalizing explanation. Perhaps this is the facet of his human temperament, separate from his scientific profile, that made him so unpopular in an intellectual era when "grand" overarching explanations presented as ultimate explanations for the natural world—from Marxism to phenomenology, from structuralism to semiotics itself, all the way to science, now presenting itself as the ultimate explanation—were all the rage:

> Totalizing interpretations are to be seen [...] with great suspicion. We can identify, in the history of culture, a number of allegedly universal keys: they are very different amongst themselves, and have not actually opened any door. Their grooves and indentations have nothing in common with any possible lock. These are systematic keys. The path to follow is precisely the opposite one: modifying our interpretive framework under the pressure of facts and interpretations, without setting the achievement of a state of ultimate knowledge as a desirable goal. This implies doubts and trials. Scientific work is intrinsically linked to the evolution of our theoretical frameworks. What is certain (that is: necessary in order to explain the mobility of these conceptual frames) is our insertion into reality as a part of it. This is the guiding presupposition for all that we can hope to achieve, and not a fideistic assumption, accepted once and for all. (Prodi 1986: 128–129)

This statement, written almost towards the end of his life (Prodi died the following year), echoes what he wrote in his very first philosophical book—*La scienza, il potere, la critica* (1974)—which therefore can be seen retroactively as a sort of programmatic manifesto of all his philosophical-scientific output that followed. *La scienza* opened with a radical critique of the concept of phenomenological *epochè*, the act of the subject's bracketing of the world, in order to reach a purely descriptive plane of apodictic and undeniable evidence. This operation, for Prodi, amounts to an impossible—both biologically and empirically—act of self-extraction from the world:

> the *epochè* is contradictory even from the point of view of its real and physical possibility, since an act that does not involve all our [existential] dimensions cannot — for the very same reason — be understood. There is nothing that can be directly revealed, infused beyond mechanisms and structures, outside of the domain of things to which mechanisms and structures belong. (Prodi 1974: 17)

In this passage we can hear the echoes of, on the one hand, Peirce's anti-Cartesian critique (see Chap. 5) and, on the other, an implicit reflection on Wittgenstein's *Tractatus* and the logical (biological as Prodi would put it) impossibility of stepping outside the world: "[p]ropositions can represent the whole of reality, but they cannot represent what they must have in common with reality in order to be able to represent it—logical form. In order to be able to represent logical form, we should be able to station ourselves with propositions somewhere outside logic, that is to say outside the world" (Wittgenstein 1922: § 4.12). Against every dualism, the human animal is nothing but a portion of the world, observing just a very limited portion of the world to which it belongs.

So far we have spoken about Prodi and his time, his proximate dialogue partners. But Prodi's thought can also be linked to another intellectual tradition: in so doing we will be able to discover how his "novelty" can really be traced back to that tradition which has recently been labelled "Italian thought" (Esposito 2012).[3] This is a tradition that has long been a marginal one, isolated from most of the history of Western philosophy which, from Descartes onwards, has been a philosophy of the "subject" and of transcendence. The Cartesian subject should be thought of as a radical and definite gesture with which the human being capable of language and self-consciousness posits itself as something different from the rest of the natural world. In this sense, Cartesian philosophy and all that followed it, all the way to phenomenology and structuralism, are philosophies of transcendence placing the subject in a position of transcendence with respect to the rest of the world. According to these traditions, the subject is beyond the world. In this sense, the philosophy of the twentieth century, the century of the "linguistic turn" (Rorty 1967), is a dualistic philosophy. The Italian philosophical tradition, on the other hand, from its origins to today—Dante, Machiavelli, Telesio, Bruno, Vico, Leopardi, Croce, and Gramsci being some of its more significant figures—has always been a philosophy of immanence. The opposite of dualism is materialism, the thesis according to which in the world there are only material entities. Materialism, in its various forms, is the contemporary philosophical fashion. If for Descartes there were two substances—*res extensa* and *res cogitans*—for the materialist, only *res extensa* really exists. As I have tried to demonstrate above, this solution is really still a prisoner of the dualist paradigm (monism is nothing but the converse of dualism, which it still presupposes and entails). The materialist holds that that dualism is in fact a material monism: a solution that has the virtue of simplicity (one axiom is better than two axioms) but that is far too simplistic. The point is that materialist monism cannot account for semiosic phenomena, i.e. the world in which *natural* "meaning" is created—what for Prodi represents, it is worth remembering, the "principal problem of philosophy" (Prodi 1989: 94). The fact is that meaning exists yet is not a thing. However, without meaning there is no biology: so the problem of materialist monism is that it cannot account for biology as a science of natural meaning. If it remains impossible to answer the question "what does it mean to say that something is meaningful?", Prodi argues, it is also impossible to understand what genetics and molecular biology are (Prodi 1979: 187). The problem of meaning, and that of semiosis, is for Prodi the acid test for any philosophical or scientific theory. Semiosis cannot be neither dualist (based on an idealist presupposition of the subject) nor grounded upon the object, like the many referentialist or correspondence theories of meaning, according to which the meaning of a sign coincides with the things to which it refers:

> Theories grounded on the assignment of meaning are sterile and idealist. Theories that take meaning and signs as primordial givens do not break out of an intransitive solipsism. It is very odd that, in this field, the idealistic stance vis-à-vis the attribution of meaning (even when disguised as a phenomenology) coincides with objectivist theories, that is to say with the acknowledgement of a categoricity intrinsic to the things that are object of examination. It can indeed be indifferently stated that a phenomenon or a thing are meaningful, in a cer-

---

[3] It is interesting to note how a similar path has been followed by the Estonian biosemiotic tradition, indicating again the significance of a specific geographical-cultural tradition (see Sebeok 1998; Tamm and Kull 2016).

tain setting, either because we say that we are, since we are giving them meaning, or because they objectively possess the meaning we discover. There is no experience that can help us discriminate between the two interpretations. There is no way out of this impasse if not by considering which relations are present, and most importantly which past, generative, relations there obtain between reality and the organism who sees it as meaningful. [...] What takes place is a natural production of meaning. (Prodi 1987a: 222–223)

On the contrary, a philosophy of immanence is a philosophy that rejects both dualism and its impoverished materialist version. A philosophy of immanence is a philosophy of nature, seen as more than the sum total of all material things. Nature is composed by all material things and all living things. A living thing is a thing, just like a cloud or a quark (assuming that quarks are things), but at the same time, it is also a *living* thing. For naturalism, the adjective "living" refers to something absolutely real linked, but not reducible, to the matter that composes it. For Prodi, a thing is alive if it partakes in meaningful relations with other things. Biology, then, is the domain of things that are meaningful for other things: this domain would not exist without matter, but to say about something that it is living is not the same as saying that it is a material entity.

Italian thought is a philosophy of immanence, and therefore an anti-dualist philosophy, but also a philosophy promoting a naturalism distinct from materialism. For this reason it has been a very peculiar philosophical tradition, not one mired in specialism (Dante and Leopardi, e.g. were poets, Vico a scholar of jurisprudence), and it never privileged scientific discourse over other kinds of discourses about the world: "Italian philosophy has always appeared poised to cross over its own boundaries; but this overstepping is precisely what allows it to achieve a perspective that would otherwise be unattainable" (Esposito 2012: 11). The vital phenomenon, precisely because it coincides with that of meaning, cannot be constrained within a single theoretical box, because life is a movement that never ceases and flourishes there where we least expect it. In this sense Italian thought, according to Esposito, is a philosophy of the *outside*, opposed to all forms of metaphysical thought: "[t]he need to "step outside" [...] rises from the difficulty they come up against when using abstract or logical-metaphysical thought to grasp something that, being effectively in motion, inevitably tends to elude them" (Esposito 2012: 11). This necessary passage towards the "outside" is clear in Prodi as well, who never stopped practicing science even while being engaged with his philosophical work, just as he never stopped being a philosopher while being a scientist: "'we are the philosophical system' would be an adequate motto" (Prodi 1983a: 11). In particular, for a philosophy of immanence, there is no separation between philosophy, strictly defined, and science, just as there is no radical separation between the material world and the world of meaning and semiosis: "the construction of theoretical functions is nothing but a way to 'follow the things as they are'" (Prodi 1983a: 12).

Let us examine closely the specific traits that, according to Esposito, characterize this intellectual tradition. It will then emerge clearly how Prodi was an *Italian* philosopher-scientist and that his timeliness is to be sought precisely in this geohistorical and intellectual specificity. Yet, a preliminary clarification is necessary: the adjective "Italian" does not refer to any alleged ethnic feature of the Italian people.

It simply indicates the belonging to Italy's peculiar cultural and civil tradition, one that extends from the era of medieval communes to the present. The first and arguably most important feature mentioned by Esposito is a persistent *"actuality of the originary"* (Esposito 2012: 23). The present time always maintains a relation with the past: history is not lost in the past, but, on the contrary, our present can exist only insofar as it is continuous with its historical past. As we have already seen, this is a crucial point for Prodi, since the human cultural world is continuous with the material world and the living world (that which is organized by "categorial logic"):

> [n]ature and culture: where is the split? There only exists a chronological succession, expressible in terms of an increase in structural complexity. In man, the natural is specialized in a cultural sense. Just as there is a linguistically-competent structure, so language — in its origin and sedimentation, in its linking the various moment of an individual with other individuals, in its weaving the before and the after of successive generations — *factually* and unitarily preserves the history of the sum total of the experiences that compose it in its shape, like a mould. Human nature contains, as a potentiality to be exploited, that which the processes that generated it, endowed it with. It is an object-testimony. It is a system for the translation of the real. It cannot be thought if not like an entity which has been generated by existing things, and this is why it is able to exercise an influence on them. Therefore, also from this point of view it is rooted in history, and it is history itself [...]. Using language amounts to using experience, accumulated over time: the structuring of linguistic competence is implicitly also a structuring of historical sense. It is rooted in us through language and it continually lives in language, even if we usually only see its synchronic side. By employing systems of communication, we necessarily employ all that contributed to their definition (Prodi 1983a: 33).

Human language can describe the world of things because it is a transformation of those very things: a radical transformation, granted, but one that nonetheless does not sever the links with the entities that it names. Language was already in the world, for otherwise it could not exist today: "the things generate their own organisms. [...] Reality lets itself be read through the production of its own readers" (Prodi 1987a: 5). In this sense, for both Prodi and Italian thought as a whole, the principle of the *"actuality of the originary"* applies. This is mainly an historical principle but also a biological one—indeed it is historical *precisely because* it is biological (and vice versa): "we cannot talk about anything if we do not know a thing: so a theory of knowledge (which does not precede but rather is within the various ways of knowing that factually develop) coincides with the identification of our origins (of our sources, phylogenesis, humanity, and of the constitution of our culture)" (Prodi 1983a: 13). But this means that the origin is not lost in the past but always again active in the present. In Italian thought, the origin is always contemporary:

> [t]his obviously has nothing to do with a mythology of the origin, by which I mean the identification of an originary moment that is identifiable as such, and from which history (or a certain kind of history) is supposed to have started and to which it could return. The genealogical attitude starts with the opposite assumption, that a founding moment of this sort is structurally absent. Because of this constitutive "inoriginarity" of history, the origin is always latently coeval with each historical moment. This allows it to be reactivated as a source of energy, rather than simply endured as some sort of spectral return. (Esposito 2012: 23)

This is a passage that Prodi could have written. His originality, however, is that in order to substantiate this idea of the contemporariness of the origin, he turned to science and biology.

However, the contemporariness of the origin means not only that the origin is active in the present but also that the present can be already found in the origin. Nature does not simply become history: nature is *already* history. A radical anti-dualism needs to be adopted, or else it would simply amount to an anodyne choice (materialist monism can hide just as much—if not more—theology than dualism can). It is not simply a matter of showing what is natural in history but also what is historical in nature. This, according to Esposito, is another distinctive character of Italian thought, the "historicization of the nonhistorical" (Esposito 2012: 26). As we have seen, this is a central point in Prodi's biosemiotic project as well. This is the domain of "material logic", both material and logic, and therefore relational, proto-semiosic:

> if we attribute to the term "logic" a meaning of "exchange or substation operation" what is really said is that such operations must exist in reality, composing the material network of facts, since our logic is nothing but a specification of these facts. From the organization of given relations (logical relations) that, in a non-contradictory way, derive more complex logical situations, both categorial and propositional (where non-contradiction is a synonym of "factual existence"). (Prodi 1982: 16)

It is only because nature is already history that it is possible to envision a continuity between first forms of relations between things and more complex ones, as in the case of animal semiosis and human language: "it is clear that material logic and material semiotic coincide" (Prodi 1977: 44). In nature there can be semiosis only because nature is already intrinsically semiosic:

> if an A interacts with B, and this interaction takes place every time they meet, this is a logi-cal operation, a material relation of a logical kind which expresses the unavoidable fabric of the real. [....] If we say that something "exists" we are referring to the operations that identify it through changes which will be both subjective — such as when every A identifies a B — and logical-connective, that is to say pertaining to the level of the uniformity and homogeneity of reality. (Prodi 1977:44)

This is an extremely original point, proper both of Prodi and of Italian thought in general. The human animal can describe the world because it is, at bottom, nothing but a piece of the world, talking about another piece of the world. *Homo sapiens* is a thing that speaks of other things; it is the world observing itself and speaking about itself (Nesteruk 2013). Fig. 12.1, borrowed from a famous essay by physicist John Wheeler, illustrates this predicament:

Both science and philosophy do not occupy a transcendent position, separated from the world they observe and describe. The philosopher and the scientist are both things that observe other things, things observing themselves: "[b]eginning with the big bang, the universe expands and cools. After eons of dynamic development it gives rise to observership. Acts of observer-participancy [...] in turn give tangible "reality" to the universe not only now but back to the beginning. To speak of the universe as a self-excited circuit is to imply once more a participatory universe" (Wheeler 1980: 362). A philosophy of immanence is a philosophy that never admits

Fig. 12.1 "The universe viewed as a self-excited circuit. Starting small (thin U at upper right), it grows (loop of U) and in time gives rise (upper left) to observer-participancy — which in turn imparts "tangible reality" [...] to even the earliest days of the universe!" (Wheeler 1980: 362)

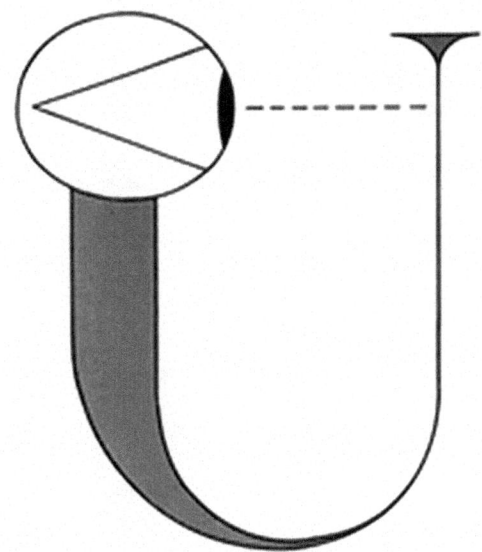

any interruption of the material and phylogenetic *continuity* between things and the human subject. Here, we once again discover the model of the circle, of the world at once observed and observing the observer, since "knowledge is always and at the same time a watching and a being watched" (Prodi 1982: 266).

The other distinctive characteristic of Italian thought, according to Esposito, is the *mundanization of the subject* (Esposito 2012: 28). As we have already seen, in the Cartesian tradition, the subject posits itself as other with respect to nature and the world:

> Not only does the modern subject presuppose itself with respect to the world of experience, it even presents itself to itself as presupposed. The subiectum suppositum, a figure posited on itself that is at the same time the substance on which it is posited, is clearly aporetic. What we have here is the construct that founds the unity of the subject on a separation between itself and its own biological substrate—or in metaphysical terms, between body and soul—crucial to the entire Western tradition. (Esposito 2012: 29)

On the other hand, Esposito argues, the Italian tradition establishes an extremely robust link between subjectivity and life, between psyche and biology: "life is not an alternative to subjectivity, but rather, constitutive of subjectivity" (Esposito 2012: 31). It should not be surprising that Prodi holds the same opinion: the human mind, as it can be seen in Fig. 12.1, is a continuation of the world of nature, taking other, and more complex, forms. Language is nothing but the most recent transformation of the "natural complementarity" of the thing A "reading" the thing B, forming the assemblage AB, and so forth. And so the world—the material world as well as the life world—is nothing but an infinitely complex tapestry of bio-semiosic relations. Language itself is the world, since "the sign is a thing". There is nothing transcendent in the subject, nothing autonomous and separate from the world, "the categories of knowledge are the phylogenesis of the reader, its historical-evolution-ary dimension with respect to the world, its capacity of reading it" (Prodi 1979: 186). The *a priori* of the single organism is really the *a posteriori* of the species to which it belongs, that is to say the result of a seamless process of mediation with the environment:

> [t]he mind did not develop in an abstract space, but in the geography and history of our bodies and relations, in the situation of inter-human relations. Human knowledge is strictly correlated to physiology in general, i.e., to the organisms' direct reactive abilities, also of an instinctual kind. If an abstracting function has developed in man, it does not mean that such abstraction should be considered as an explanation: it rather should be naturalistically explained, as a goal and not a point of origin. It remains in the immanent domain of rela-tions and it cannot be disguised as a supernatural principle, linked to non-natural essences or schemas. The fundamental mistake of anthropocentrism was that of identifying the abil-ity for abstract thought with real "abstract entities". Moreover, knowledge remains pro-foundly psychological, that is to say, linked to both the situation of those mechanisms that make it explicit and the single individual in which it occurs. (Prodi 1979: 184–185)

The *mundanization of the subject*, for Prodi, means that the subject does not stop being a thing, even when it comes to know the world and, most of all, when she knows itself. There is no knowledge without contact with the environment, and this starts with the contact with one's "own" body, since all knowledge is always situ-ated (Gibson 1979; Kirshner and Whitson 1997) and never disincarnate and abstract: "consciousness is the location of one's placement as an individual, both in the flux of history and in its geographical dimension. It is the capacity to evaluate oneself in relation to their external correlates, individuated via natural-cultural means" (Prodi 1979: 185). The subject, even when conceived as pure self-consciousness, never stops being a thing of the world. If it was not, it could not know nothing since "to know a thing is to be changed by it. Knowledge is always, at every level, a process of the change of things undergoing reciprocal adaptation" (Prodi 1979: 185). Fig. 12.1, at bottom, is another version of Fig. 8.1, the one illustrating the Möbius strip: the subject is something that can be observed; it is an introspection of the natu-ral world. The three distinctive characteristics of Italian thought, as listed by Esposito—"actuality of the originary", "historicization of the nonhistorical", and "mundanization of the subject"—all constitute central points of Prodi's thought. In this sense, he was an Italian naturalist philosopher, since this was the tradition within which his thought developed. According to Prodi, coherently with this tradi-tion but very differently from much contemporary philosophy (even parts of Italian philosophy!), the project is not one of substituting philosophy with science nor keeping philosophy in an a-historic situation of isolation. On the one hand, it is necessary to show how science is linked to natural semiosis, deriving from "mate-rial logic" and therefore unable to stand on any higher ground with respect to other forms of knowledge (science is not transcendent with respect to the world); on the other, it is also necessary to show how philosophy is first of all a reflection on *mean-ing* and therefore on how this is constituted in nature.[4] Science knows the world,

---

[4]A history of the "materialist" Italian tradition, extending to literature as well, can be found in Antonello (2012). Regarding semiotics in particular, we could also mention the work of Rossi

philosophy knows the knowledge about the world. Both are necessary forms of knowledge, and both derive directly form the world:

> [V]ico [...] used to say that one can only know what one does, and that man can only know history, since that is what he can do. On these grounds, he argued for a strict separation between historical sciences (true knowledge) and enquiry into nature. [...] I want to say that Vico was right, but that history is natural history, which makes us what we are, and that we can know nature only because it made us (or we made ourselves within it). It is for this reason that there is no dichotomy within knowledge. And in this perspective, only an adequate philosophy of knowledge (and of scientific knowledge) can recuperate and absorb the entire tradition, i.e., the historical essence of man, and its various disciplines. For this reason too, it is legitimate to consider scientific research as natural hermeneutics. [...] When it comes to the history of philosophy, we go back to the turn between the pre-Socratics and Plato. Before Socrates, they used to talk about *nature* and *generation*. After Socrates, mostly about logic and deduction. This was a shift from natural facts to Mind. Now we can go back to the Pre-Socratics, and therefore once again speak of nature and generation. These, however, now include logic and deduction: demonstrating that these, in all their rigour, were generated by nature. Plato is reabsorbed into physics. (Prodi 1987b: 119)

Landi (Rossi 1992) and Augusto Ponzio (Deely et al. 2005), as well as Barbieri's (2015) "Code Biology" as belonging—albeit indirectly—to "Italian thought". Barbieri's case, however, is an interesting one, arguably being the closest to Prodi's original project. Indeed, Prodi was a scientist who never renounced the scientific approach which is, in some measure, always mechanistic: "the existence of organic codes and organic meaning in nature are scientific problems that can and should be investigated with the classical method of science, i.e. the mechanistic approach of model building" (Barbieri 2014: 241). The crucial point is the relation with Peirce's semiotics. We have seen how, for Prodi, the sign coincides with the referent in the initial semiosis (the semiotic triangle is flattened to a line segment). So, in this case, there is no *interpreter* mediating between sign and referent. On the contrary, contemporary biosemiotics thinks of "decoding as a form of interpretation" (Barbieri 2015: 245). Barbieri is a critic of this identification, since the concept of "interpretation" is not scientifically—i.e. mechanistically—definable. Consequently, Barbieri concludes that "a scientific approach to the semiosis of Nature could not prosper within that framework, and…its future was seriously at risk" (Barbieri 2015: 246). It should not be forgotten that, according to Prodi, there was a "metaphysical Peirce, who transformed all semiotic matter into something spiritual" (Prodi 1986: 126), and that Prodi's approach to natural semiosis is precisely the opposite: "it is necessary to identify, and recognize ourselves in, a 'natural rationalism' which is different from the rationalism of geometry. It does not proceed from our logic towards the things, but takes the opposite path" (Prodi 1986: 126).

Biosemiotics by Giorgio Prodi: A Postscript

Kalevi Kull

Giorgio Prodi was a co-founder of the biosemiotic field—the biosemiotic science or the contemporary interdiscipline of biosemiotics[1]—due to his works from the 1970s and 1980s. Prodi wrote much, but almost only in Italian. Thomas Sebeok dedicated the first volume ever published under the title *Biosemiotics* to him, with the inscription "*In memoriam*: Giorgio Prodi (1928–1987): bold trailblazer of contemporary biosemiotics" (Sebeok and Umiker-Sebeok 1992: v). Donald Favareau included Prodi in his list of 24 essential authors of biosemiotics in the first ever anthology of biosemiotics (Favareau 2010a, b; Prodi 2010a). Thus, Prodi is a classic in the field. However, an in-depth analysis of his work from a biosemiotic point of view has hitherto still been absent.

So it was obvious, after learning that Felice Cimatti has written both about zoosemiotics (Cimatti 1998) and about Prodi (Cimatti 2000a) in Italian, to ask him to publish an article on Prodi's biosemiotics in English (Cimatti 2000b) and, since he had published a whole book about Prodi in Italian, to publish also an updated version in English—which is what we have here.[2]

Prodi was a very productive writer. The list of his publications (Prodi 1987) divides his works into two categories. The category "scientific" includes 10 books (all in Italian, except 1 edited volume of conference proceedings in English) and 324 articles[3] (most of these with co-authors, 178 in English, a few in French, the rest in Italian). The category "philosophical and literary" includes 13 books (all in

---

[1] Favareu 2010a: 280.

[2] I am thankful for the brief but productive meetings with Felice Cimatti in Rome (April 2002), San Marino (June 2002), and Palermo (December 2016). On recent zoosemiotic (or ecosemiotic) work by him, see also Cimatti (2018). I also thank The University of Tartu for support via IUT2-44.

[3] The first among these published in 1953.

Kalevi Kull
University of Tartu, Tartu, Estonia
e-mail: Kalevi.Kull@ut.ee

© Springer Nature Switzerland AG 2018
F. Cimatti, *A Biosemiotic Ontology*, Biosemiotics 18,
https://doi.org/10.1007/978-3-319-97903-8

Italian) and 66 articles (most of these single-authored, 6 in English, the rest in Italian). Works on semiotics are all listed in the latter category.

Prodi's first article on semiotics was "The prehistory of sign" (Prodi 1974).[4] The (hopefully) complete list of Prodi's works on semiotics which exist in English is the following:

(1) "Material bases of signification", published in *Semiotica* (Prodi 1988a). An abbreviated translation of Prodi 1977.
(2) "Development of semiosic competence", published in the *Encyclopedic Dictionary of Semiotics* (Prodi 1986a; reprinted in Prodi 1994 and Prodi 2010b). First version of this text—with preliminary titles "Phylogeny of codes" and "Ontogeny of codes"—was written in 1981 (as mentioned in Prodi 1987: 60).
(3) "Biology as natural semiotics" (Prodi 1989c), published in a Bochum semiotics series, was initially a talk given at the Third Congress of the International Association for Semiotic Studies in 1984 in Palermo. An Italian version of this article has been published repeatedly (Prodi 1984, 1988e, 2002).[5]
(4) "Culture as natural hermeneutics" (Prodi 1989b), published in the proceedings volume of the 1986 Bochum conference *The Nature of Culture*. An Italian version published in Prodi 1988d.
(5) "Signs and codes in immunology", published in the proceedings volume of the 1986 Lucca conference *The Semiotics of Cellular Communication in the Immune System* (Prodi 1988c). Reprinted in *Essential Readings in Biosemiotics* (Prodi 2010a).
(6) "Toward a biologically grounded ethics" in the Bologna University publication *Alma Mater Studiorum* (Prodi 1989d) is printed together with the Italian text, which was a talk given at the 1987 conference "Ethics of scientific knowledge" in Venice.

As Prodi's work on biosemiotics was highly evaluated by Thomas Sebeok and Umberto Eco, let us take a brief look at their reflections on Prodi's semiotic thought and point at some relevant work and recent ideas that are close to Prodi's approach.

From Thomas Sebeok

When speaking about the development of fields in academic endeavours, Thomas Sebeok often referred to Mihály Csíkszentmihályi, a psychology professor from the University of Chicago, who made a distinction between domains and fields (e.g. Sebeok 2004). Every scholarly field has its gatekeepers, they both said, who take responsibility for the limits as well as the purity (or rather productivity and

---

[4] See also an interview with Prodi (Prodi 1986b).

[5] Prodi attended also the Second Congress of the International Association for Semiotic Studies in 1979 in Vienna, with a talk "The origin of meaning in phylogenesis" (as mentioned in Prodi 1983: 202), but his paper was not published in the congress proceedings (1984, edited by Tasso Borbé).

creativity) of the field, letting some work or some researchers in and leaving others out as inappropriate for reasons such as quality, etc. Indeed, this is what the leading figures in academic fields often do.

However, what is more important than gatekeeping and what our mentors should be praised for is window-opening. Gatekeeping is a negative function, while the window-opening is a positive and supportive one.

The windows of fitting (recognition windows) are important instruments in how science works. Whom to follow, whom to take in or to take with, whom to mention just briefly, and whose ideas rather to avoid—each scholar has his or her windows: the windows of limits of knowledge and of hopes in understanding of study objects. The mentors' windows are those that lead to groundbreaking paths in research, innovative research programmes.

Scholars in the field of biosemiotics would describe their domain as covering everything related to the primary mechanisms of meaning-making or, from another aspect, to the mechanisms of embodiment.[6] Biosemiotics deals with the meaning-making that is real in non-human organisms. The study field of the primary meaning-making has appeared stepwise, and it was Thomas Sebeok who introduced to us Giorgio Prodi as a pioneer of biosemiotics, mentioning him in several talks.

Prodi incorporated semiotics into biology just at the time when Sebeok rediscovered Jakob von Uexküll. Sebeok paid attention to the fact that there was more in biology that belonged to the domain of semiotics than animal behaviour as studied by ethology. This happened in the late 1970s. Yet there were only a few researchers in biosemiotics at that time.

Sebeok was famous for his ability to connect scholars. In a couple of his articles, he recalls a week spent in the company of Thure von Uexküll and Giorgio Prodi in Freiburg in 1979. For instance (Sebeok 2004: 87–88):

> Thure made arrangements for me to spend a week or so visiting him in Freiburg. [...] Our Freiburg discussions about multifarious biosemiotic topics were carried out, with rare intensity, from morning late into every night, and were happily augmented by the continuous participation of Giorgio Prodi, Director of the Institute for Cancer Research of the University of Bologna. Prodi, an astounding polymath [Eco 1988b] who had become my friend several years earlier, encountered Thure for the first time on that occasion; the two of them met only twice more, first in Palermo in the Summer of 1984, then the last time in Lucca in the early fall of 1986. [Sercarz et al. 1988]

He offers some additional details of that period (Sebeok 2004: 91):

> [...] during the week we spent together in our open-ended 1979 'intensive seminar' in Thure's company in Freiburg on the practical and conceivable ins and outs of biosemiotics, the three of us got along extremely well; as I commented afterwards, 'this uniquely stimulating experience enabled me to enhance my writings and teachings [...] in biosemiotics in its various topical subdivisions'. [Sebeok 1998: 34–35]

However, this does not mean that Sebeok would agree with all of Prodi's views on biosemiotics. For instance, Sebeok was critical of some terminology used by Prodi (Sebeok 1997a: 436):

---

[6] Cf. Emmeche (2007: 385): "biosemiotics—a qualitative organicist account of embodiment".

The quasisemiotic phenomena of nonbiological atomic interactions and, later, those of inorganic molecules, were consigned by the late oncologist Prodi (1977) to 'proto-semiotics', but this must surely be read as a metaphorical expression. Prodi's term is to be distinguished from the notion 'primitive communication', which refers to the transfer of information-carrying endoparticles, such as exists in neuron assemblies, where it is managed in modern cells by protein particles.

Sebeok did not accept Prodi's term "natural semiotics" ("semiotica naturale", Prodi 1988b: 149), which he said to be a poor substitute for "biosemiotics" (Sebeok 2004: 90).

In Sebeok's library, which now belongs to the Semiotics Department of the University of Tartu, there are ten books by Prodi, plus a set of reprints of his articles on semiotics.

From Umberto Eco

In our conversation with Umberto Eco in Milan in 2012, he mentioned: "When I discovered the research of Giorgio Prodi on biosemiotics I was the one who published his book that maybe I was not totally agreeing with, but I found it was absolutely important to speak about those things". Two books by Prodi (Prodi 1979; Prodi 1982) appeared in the series edited by Eco (Espresso Strumenti, and Studi Bompiani: Il campo semiotico).

Describing Prodi's impact, Eco pointed out a major aim of biosemiotics (Eco 2004: 27–28):

> [...] the assumption that both a genetic and an immunological code can in some sense be analysed semiotically seems to constitute the new scientific attempt to find a language that can be defined as a primitive par excellence, though not in historical, but rather in biological terms. This language would rest in the roots of evolution itself, stretching back to before the dawn of humanity. This was the thesis of our friend Giorgio Prodi[7], published by Tom Sebeok even in English[8]. Just one remark, that in this last case Prodi was not looking for the historical origins of language, but rather for the biological roots of semiosis, which is a different approach.

Thomas Sebeok, in his foreword to a book about Umberto Eco, devotes a surprisingly lengthy passage to Giorgio Prodi (Sebeok 1997b: xiv):

> I do believe it is appropriate for me to note here [his – Umberto Eco's] involvement with yet another among our mutual friends, the late Giorgio Prodi (1928–1987), Eco's near-contemporary colleague at the University of Bologna, and himself the scion of a very distinguished family of Italian public servants and academics, severally close to Eco. Prodi was a prodigiously busy polymath, in some way out-Ecoing Eco: 'Perché [Giorgio] aveva una giornata di quarantott'ore e noi di sole ventiquattro?',[9] Umberto questioned in mock-peeve. Indeed, Prodi was, on the one hand, one of his country's leading medical biologists

---

[7] Reference to Prodi 1977

[8] Prodi 1988a; that is a partial translation of Prodi 1977

[9] "Why [Giorgio] had a day of forty-eight hours and we one of only twenty-four?" (in Italian).

in oncology, while he was, on the other, a highly original contributor to semiotics and epis-temology, the philosophy of language and formal logic, plus a noteworthy literary figure. An immensely prolific scientist, Prodi was one of a handful of European pioneers in the exploding transdisciplinary field that has come lately to be dubbed biosemiotics. The year before Prodi died, he and Eco together took part in a landmark meeting in Lucca,[10] juxta-posing semiotics and immunology, bringing the two, as it were, under a new interdisciplin-ary branch of biological sciences, 'immunosemiotics', which is now, with a different emphasis, an important branch of biosemiotics. Prodi's earliest contribution to this area, *Le basi materiali della significazione*, was published first in Eco's journal *Vs* (Prodi 1976),[11] then boldly the following year in one of the well-known Bompiani series also edited by him.[12] Again, his beautiful, characteristically informed and observant *Ricordo* of Prodi's life and accomplishments, 'Una sfida al mito delle due culture' [Eco 1988b], repays close study for what it tells us about Giorgio no less, to be sure, for what it reveals about Umberto.

Indeed, Eco dedicated a couple of his talks to Prodi, speaking about "A challenge to the myth of the two cultures" (Eco 1988b, 1989; in English translation, Eco 1994) and "Giorgio Prodi and the lower threshold of semiotics" (Eco 1988c).

In *Semiotics and the Philosophy of Language*, when discussing the genetic code as a possible example of s-code, Eco refers to Prodi's (1977) position that genetic code "represents an elementary, but by no means metaphorical, example of *inter-pretation* in Peirce's sense" (Eco 1986: 184).

Prodi was a biologist whom Eco trusted and whose work in biosemiotics he took seriously. Eco told about this in a talk from 1988, devoted to Prodi (Eco 1988c: 4–5):

> When, in 1974, Giorgio Prodi visited me for the first time, I was still very suspicious towards every semiotic approach to the cellular universe. But I was struck by the Copernican revolution that I glimpsed in his talk. For the first time, I was facing a scientist who was not telling me 'maybe cells speak like us', but, rather, 'maybe we speak like cells'. In this simple inversion of the terms of the problem lies the originality of Prodi's approach.

As Eco describes Prodi's conclusions, he repeats upon reading Prodi's last article (Prodi 1988d; in English Prodi 1989b): "hermeneutics is not a late product of cul-ture, but the same elementary movement of life, that is born because something obscurely interprets something else" (Eco 1988c: 8). Eco adds that Prodi's "pro-posal remains a challenge that I do not think has been welcomed yet in all its impli-cations" (Eco 1988c: 8).

Moreover, in *Kant and the Platypus*, Umberto Eco introduces an explicit theo-retical connection between his searches and Prodi's (Eco 1999: 107):

> I am admitting with Prodi (1977) that to understand the higher cultural phenomena, which clearly do not spring from nothing, it is necessary to assume that certain 'material bases of signification' exist, and that these bases lie precisely in this disposition to meet and interact that we can see as the first manifestation (not yet cognitive and certainly not mental) of primary iconism.

---

[10] See Eco 1988a and Prodi 1988c.

[11] Another Prodi's article in *Versus*—"Interpretation as a change of the interpreter"—was published posthumously in Prodi 1989a.

[12] Prodi 1977

That last point has already been commented on by Cimatti. As I feel closeness to all three, Peirce, Eco, and Prodi (with some preference for Eco's arguments), let us add some comments on the concepts and phenomena of the "proto-semiotic" realm.

Earlier and Later: "Simple" Semiotic Processes and Structures

To what extent, in what sense, is meaning-making relevant in brainless, even non-animal organisms? Biosemiotics offers a series of approaches to deal with this. Some have been mentioned by Prodi—Charles Peirce, Jakob von Uexküll, Charles Morris, Jean Piaget, Peter Marler, W. John Smith,[13] and Thomas A. Sebeok (Prodi 1983, 1988b). However, Prodi does not discuss explicitly the other approaches, already existing in biosemiotics by the 1980s, and he could not discuss several ones that have appeared since 1987.

Eco warns us: "With all due caution: in no way am I repudiating the distinction (which remains fundamental) between signal and sign, between dyadic processes of stimulus-response and triadic processes of interpretation, so that only in the full expansion of this last do phenomena such as signification, intentionality, and interpretation (however you wish to consider them) emerge" (Eco 1999: 107).

Accepting that sign process, semiosis, requires triadicity, then in what sense, if at all, can one speak about dyadic processes as related to semiosic ones? Is it not incorrect to speak about "the natural willingness of something to *correspond to* something else" (Eco 1999: 107) at that level? What exactly is meant by triadicity as an irreducible feature of interpretation, of sign?

The origin of irreducible triadicity, or the necessary and sufficient conditions for semiosis, has been widely studied and discussed in semiotics (see, e.g. Alač and Violi 2004; Nöth 1994; Merrell 2013; etc.). In the field of biosemiotics, these questions have received a whole variety of proposed answers (Hoffmeyer 1996, 2009; Barbieri 2007; Emmeche and Kull 2011; Maran et al. 2011; etc.). Let us mention some that look closest to Prodi's.

Marcel Florkin, a biochemist, attempted to apply Saussurean-Greimassian terminology and approach to the cellular processes of recognition. He states, somewhat similarly to Prodi: "In molecular biosemiotics, [...] significant and signified are in a necessary relation imposed by the natural relations of material realities" (Florkin 1974: 14). In the semiotic theory of Saussure and Greimas, the problem of triadicity does not exist explicitly.

For Hoffmeyer and Claus (1991), on the other hand, after the application of Peircean model and the concept of code-duality, the problem of proto-semiosis does not appear, for, as long as there is semiosis—already on the cellular level, as they

---

[13] Smith 1965

state—it is triadic, and there is interpretation. Frederik Stjernfelt (2014) has expressed a similar view.

A researcher who applied semiotic concepts on the cellular level somewhat similar to Prodi was the Japanese biochemist Yoshimi Kawade. One of his first articles on biosemiotics (Kawade 1992) appeared in the Italian journal *Rivista di Biologia*— the journal that for a long time was edited by Giuseppe Sermonti, a structuralist biologist of non-Darwinian views and a member of the Osaka group.

There exist attempts to formalize the concept of interpretation, illustrated by an example of an amoeba crawling along a chemical gradient that leads to a source of food (Robinson and Southgate 2010). Using this operational definition of interpretation, Lehman et al. (2014) demonstrated its applicability for the catalytic behaviour of RNA.

On the other hand, the Italian biologist Marcello Barbieri does not accept the usage of the concept of interpretation for nonanimal organisms. He focuses on the description of codes (Barbieri 2000, 2001). The feature, characteristic of codes, but simpler than interpretation, is mediatedness: for instance, the genetic code is both made and mediated by tRNAs and ribosomes (Barbieri 1985).

More recently, Prodi's term "proto-semiosis"[14] has been picked up by Aleksei Sharov and Tommi Vehkavaara (Sharov 2013, 2017; Sharov and Vehkavaara 2015). They proposed a redefinition of proto-semiosis "as a kind of sign processing, where agents (i.e. active systems guided by natural self-interest) initiate or modify their functional activities in response to incoming signs directly, rather than by associating signs with objects. [...] Proto-semiosis is opposed here to 'eusemiosis' where signs are associated with objects and interpretants" (Sharov and Vehkavaara 2015: 107–108).

At a recent meeting of the Tartu biosemiotics research seminar (April 2018), we discussed the typology of incomplete signs. Tyler Bennett proposed the term *tardosigns* to be used for *post-signs* as opposed to *proto-signs*, which leads to a possible typology of "as if" dyadic signs. Thus, signs without objects are signals, or proto-signs as Sharov and Vehkavaara have argued, while signs without interpretant are tardosigns, and signs without representamen could be symptoms. Indeed, symptoms per se, as these appear in an organism itself, directly connect the object with its interpretant. (A symptom as a sign for a physician should be analysed differently.) Yet while it may indeed be instructive to understand certain differences in these relations,[15] I do not think this speculation with terms should be a proposal for a new terminology.

Nevertheless, a fundamental question still remains about the difference between semiosis and non-semiosis or signs and non-signs. This question is not removed by the distinction between proto-signs and eu-signs, as the usage of the term proto-sign or proto-semiosis is justified only if proto-semiosis is semiosis in some sense. A solution proposed for this problem turns attention to the interdependent and

---

[14] The term was used in Prodi (1988c: 55) while not yet in Prodi (1988a [1977]).

[15] In addition, see Sebeok (2001: 75) on the concepts of *protosemiosic*, *microsemiotic*, etc; Petrilli, Ponzio (2005: 207) on *quasisemiotic*; and Deely (2009: 111ff) on *physiosemiotic*.

simultaneous existence of possibilities and the process of choice. Interpretation (which is semiosis, according to Peirce) assumes the choice between possibilities, i.e. optionality.[16] We would define an (semiotic) agent as a system with spontaneous activity that chooses its actions. For a choice to exist, the options should be simultaneous; otherwise there cannot be any choice. The latter means that semiosis is coextensive with the phenomenal present (Kull 2015, 2017, 2018). The demonstration of early (and distributed) forms of phenomenal present will be a fascinating task.

A choice that an interpretation makes is the mechanism that introduces semiotic order in the form of habits or codes—i.e. the local relations—produced by semiosis. Habits and codes working automatically can be described as incomplete semiosis, or relations without semiosis. However, as products of semiosis, such relations can be (and have been) a research object for semiotics. Only if proto-signs as described by Prodi and others are products of earlier triadic semiosis is it justified to apply semiotic models to describe them. A careful analysis of ontological assumptions made by semioticians, including Felice Cimatti in the current book, is an important task in biosemiotic research, helping to discover the scope of applicability of semiotics.

Conclusions

The philosophical basis, the ontology, is certainly fairly dissimilar in case of different biosemioticians: compare, for instance, the conceptual bases in Uexküll (1928), Hoffmeyer (2009), Deacon (2011), Barbieri (2001), Weber (2016), Pattee (Pattee and Rączaszek-Leonardi 2012), and Prodi (1988a). The ontology of meaning, of sign, and of semiosis is what largely determines the whole of the semiotic theory one can build.[17] The major discussions—and misunderstandings—in contemporary general semiotics (i.e. by no means only in biosemiotics) are stemming largely from the differences in the ontologies used by scholars. Either there is a (lower) semiotic threshold or there is not; if there is, the question remains where (Eco 1988c; Rodríguez Higuera and Kull 2017) and whether the acceptance of the reality of the lower semiotic threshold would mean the acceptance of Cartesianism or not, i.e. whether the dualistic concepts are different from the dual ones.

Ontology also has implications for methodology. This can be seen from the debates about methods and scientificity between the scholars who limit semiotic

---

[16] The existence of choice and options as necessary conditions for interpretations and semiosis is also used by Umberto Eco as an argument for the limits of the domain of semiotics (e.g. Eco 1988c).

[17] See, for instance, Emmeche (1999), Tønnessen (2001), Buchanan (2008), Kockelman (2013), Rodríguez Higuera (2016), Bardini (2017), and Bárdos and Zemplén (2017) for some explicit analyses of biosemiotic ontologies.

research with the study of codes and others who find the main object of semiotics to be interpretation (see Markoš 2010; Cobley 2016; Favareau et al. 2017).

In contemporary Italy, courses in semiotics are taught at almost every university, while the role of biosemiotics in these is small. Considering the textbooks used (see Kull et al. 2015), or the approaches in zoosemiotics that prevail in Italy (Marrone and Mangano 2018), Prodi's approach may look rather exceptional. In this context, it is interesting to see how Cimatti describes the relationship of Prodi's philosophy to Italian theory (or Italian thought) as this concept has recently been explicated (Esposito 2015; Gentili 2012; Claverini 2016). Indeed, the understanding of a certain broad domain, and the ways in which the understanding is expressed, bears a remarkable local cultural trace.[18]

Italian semiotics is a rich part of contemporary semiotics that is witnessing much exciting activity. The analysis of primary forms of semiosis has been a remarkable good part of it, as demonstrated above. The origin of meaning-making is a difficult problem to solve, and a high-level scholarly tradition is necessary for this. We have reason to believe that Italian forums in semiotics will lead to further discoveries in the biosemiotic domain.

References

Alač, M., & Violi, P. (Eds.). (2004). *In the beginning: Origins of semiosis, Semiotic and Cognitive Studies 12*. Bologna: Brepols.

Barbieri, M. (1985). *La teoria semantica dell'evoluzione*. (Thom, René, preface.) (Saggi, Scienze.) Torino: Bollati Boringhieri.

Barbieri, M. (2000). *I codici organici: La nascita della biologia semantica*. (Capire la vita 1.) Ancona: peQuod editore.

Barbieri, M. (2001). *The organic codes: The birth of semantic biology*. (Capire la vita 2.) Ancona: peQuod editore.

Barbieri, M. (Ed.). (2007). *Introduction to Biosemiotics: The New Biological Synthesis*. Berlin: Springer.

Bardini, T. (2017). Relational ontology, Simondon, and the hope for a third culture inside biosemiotics. *Biosemiotics, 10*(1), 131–137.

Bárdos, D., & Zemplén, G. Á. (2017). The shape of biology to come? The account of form and form of account in Hoffmeyer's biosemiotics. *Tradition & Discovery: The Journal of the Polanyi Society, 43*(1), 32–50.

Buchanan, B. (2008). *Onto-Ethologies: The animal environments of Uexküll, Heidegger, Merleau-Ponty, and Deleuze*. Albany: State University of New York Press.

Cimatti, F. (1998). *Mente e linguaggio negli animali: Introduzione alla zoosemiotica cognitiva*. Roma: Carocci editore.

---

[18] Sharing preferences locally does not require personal influence between scholars. For Instance, as Marcello Barbieri informed me in a letter (6th of May 2018), Prodi's Institute was only two blocks away from his, and while Barbieri did have a few discussions with Prodi, these were only concerned with the molecular basis of cancer. Barbieri learned of Prodi's interest in biosemiotics only from Sebeok, and it was a major surprise for him. Barbieri says that their approaches to biosemiotics were totally independent of each other.

Cimatti, F. (2000a). *Nel segno del cerchio: L'ontologia semiotica di Giorgio Prodi*. Roma: Il manifesto Libri.
Cimatti, F. (2000b). The circular semiosis of Giorgio Prodi. *Sign Systems Studies, 28*, 351–379.
Cimatti, F. (2018). *Sguardi animali*. Milano: Mimesis.
Claverini, C. (2016). La loso a italiana come problema: Da Bertrando Spaventa all'Italian Theory. *Giornale Critico di Storia delle Idee, 15*(16), 179–188.
Cobley, P. (2016). *Cultural Implications of Biosemiotics, Biosemiotics 15*. Dordrecht: Springer.
Deacon, T. W. (2011). *Incomplete Nature: How Mind Emerged from Matter*. New York: W. W. Norton & Co..
Deely, J. (2009). *Basics of Semiotics, Tartu Semiotics Library 4.2* (5th ed.). Tartu: Tartu University Press.
Eco, U. (1986 [1984]). *Semiotics and the philosophy of language*. Bloomington: Indiana University Press.
Eco, U. (1988a). On semiotics and immunology. In: Sercarz et al. 1988: 3–15.
Eco, U. (1988b). Una sfida al mito delle due culture. In: *Saecularia nona*, 2. [In English, Eco 1994.]
Eco, U. (1988c). Giorgio Prodi e la soglia inferiore della semiotica. [Giorgio Prodi and the lower threshold of semiotics.] (Manuscript in T. A. Sebeok memorial library, Department of Semiotics, University of Tartu, Estonia.) [In English in *Sign Systems Studies* 46(2/3), 2018.]
Eco, U. (1989). Una sfida al mito delle due culture. *Il Belpaese 7*: 166–168. [Reprint of Eco 1988b.]
Eco, U. (1994). In memory of Giorgio Prodi: A challenge to the myth of two cultures. (Johnston, Marina, trans.) In: Jaworski, L. G (ed.), *Lo studio bolognese: campi di studio, di insegnamento, di recerca, di divulgazione* (pp. 75–78). Stony Brook: Forum Italicum (Center for Italian Studies, State University of New York at Stony Brook), [Translation of Eco 1988b.]
Eco, U. (1999 [1997]). *Kant and the platypus: Essays on language and cognition*. (A. McEwen, Trans.). San Diego: A Harvest Book, Harcourt.
Eco, U. (2004). Origins of semiosis. In: Alač, Violi 2004: 25–30.
Emmeche, C. (1999). The biosemiotics of emergent properties in a pluralist ontology. In E. Taborsky (Ed.), *Semiosis, evolution, energy: Towards a reconceptualization of the sign* (pp. 89–108). Shaker Verlag: Aachen.
Emmeche, C. (2007). On the biosemiotics of embodiment and our human cyborg nature. In T. Ziemke, J. Zlatev, & R. M. Frank (Eds.), *Body, language and mind. Vol. 1: Embodiment, Cognitive Linguistics Research 35.1* (pp. 379–410). Berlin: Mouton de Gruyter.
Emmeche, C., & Kull, K. (Eds.). (2011). *Towards a Semiotic Biology: Life is the Action of Signs*. London: Imperial College Press.
Esposito, R. (2015). *German philosophy, French theory, Italian thought*. In D. Gentili & E. Stimilli (Eds.), *Differenze italiane. Politica e loso a: mappe e scon namenti*. Roma: Derive Approdi.
Favareau, D. (Ed.). (2010a). *Essential readings in biosemiotics: Anthology and commentary, Biosemiotics 3*. Berlin: Springer.
Favareau, D. (2010b). Introduction and commentary: Giorgio Prodi (1928–1987). In: Favareau 2010a: 323–327.
Favareau, D., Kull, K., Ostdiek, G., Maran, T., Westling, L., Cobley, P., Stjernfelt, F., Anderson, M., Tønnessen, M., & Wheeler, W. (2017). How can the study of the humanities inform the study of biosemiotics? *Biosemiotics, 10*(1), 9–31.
Florkin, M. (1974). Concepts of molecular biosemiotics and of molecular evolution. *Comprehensive Biochemistry, 29A*, 1–124.
Gentili, D. (2012). *Italian theory: Dall'operaismo alla biopolitica*. Bologna: Il Mulino.
Hoffmeyer, J. (1996). *Signs of meaning in the universe*. Bloomington: Indiana University Press.
Hoffmeyer, J. (2009). *Biosemiotics: An examination into the signs of life and the life of signs*. Scranton: University of Scranton Press.

Hoffmeyer, J., & Claus, E. (1991). Code-duality and the semiotics of nature. In M. Anderson & F. Merrell (Eds.), *On semiotic modeling, Approaches to Semiotics 97* (pp. 117–166). Berlin: Mouton de Gruyter.

Kawade, Y. (1992). A molecular semiotic view of biology: Interferon and 'homeokine' as symbols. *Rivista di Biologia – Biology Forum, 85*(1), 71–78.

Kockelman, P. (2013). *Agent, person, subject, self: A theory of ontology, interaction, and infrastructure.* Oxford: Oxford University Press.

Kull, K. (2015). Semiosis stems from logical incompatibility in organic nature: Why biophysics does not see meaning, while biosemiotics does. *Progress in Biophysics and Molecular Biology, 119*(3), 616–621.

Kull, K. (2017). On the limits of semiotics, or the thresholds of/in knowing. In T. Thellefsen & B. Sørensen (Eds.), *Umberto eco in his own words, Semiotics, Communication and Cognition 19* (pp. 41–47). Berlin: De Gruyter Mouton.

Kull, K. (2018). On the logic of animal umwelten: The animal subjective present and zoosemiotics of choice and learning. In M. Gianfranco & D. Mangano (Eds.), *Semiotics of animals in culture: Zoosemiotics 2.0, Biosemiotics 17* (pp. 135–148). Cham: Springer.

Kull, K., Bogdanova, O., Gramigna, R., Heinapuu, O., Lepik, E., Lindström, K., Magnus, R., Moss, R. T., Ojamaa, M., Pern, T., Põhjala, P., Pärn, K., Raudmäe, K., Remm, T., Salupere, S., Soovik, E.-R., Sõukand, R., Tønnessen, M., & Väli, K. (2015). A hundred introductions to semiotics, for a million students: Survey of semiotics textbooks and primers in the world. *Sign Systems Studies, 43*(2/3), 281–346.

Lehman, N., Bernhard, T., Larson, B. C., Robinson, A., & Southgate, C. (2014). Empirical demonstration of environmental sensing in catalytic RNA: evolution of interpretive behavior at the origins of life. *BioMed Central Evolutionary Biology, 14*, 248.

Maran, T., Martinelli, D., & Turovski, A. (Eds.). (2011). *Readings in zoosemiotics, Semiotics, Communication and Cognition 8.* Berlin: De Gruyter Mouton.

Markoš, A. (2010). Biosemiotics and the collision of modernism with postmodernity. *Cognitio, 11*(1), 69–78.

Marrone, G., & Mangano, D. (Eds.). (2018). *Semiotics of animals in culture: Zoosemiotics 2.0, biosemiotics 17.* Cham: Springer.

Merrell, F. (2013). *Meaning making: It's what we do; It's who we are, Tartu Semiotics Library 12.* Tartu: University of Tartu Press.

Nöth, W. (Ed.). (1994). *Origins of semiosis: Sign evolution in nature and culture, Approaches to Semiotics 116.* Berlin: Mouton de Gruyter.

Pattee, H. H., & Rączaszek-Leonardi, J. (2012). *Laws, language and life: Howard Pattee's classic papers on the physics of symbols with contemporary commentary, Biosemiotics 7.* Berlin: Springer.

Petrilli, S., & Ponzio, A. (2005). *Semiotics unbounded: Interpretive routes through the open network of signs.* Toronto: University of Toronto Press.

Prodi, G. (1974). La preistoria del segno. *Lingua e Stile, 9*(1), 117–145.

Prodi, G. (1976). Le basi materiali della significazione. *Versus, 13*, 69–93.

Prodi, G. (1977). *Le basi materiali della significazione, Nuovi saggi italiani 21.* Milano: Bompiani.

Prodi, G. (1979). *Orizzonti della genetica.* (Espresso Strumenti 6, a cura di Umberto Eco.) Editoriale l'Espresso.

Prodi, G. (1982). *La storia naturale della logica.* (Studi Bompiani: Il campo semiotico, a cura di Umberto Eco.) Milano: Bompiani.

Prodi, G. (1983). Linguistica e biologia. In C. Segre (Ed.), *Intorno alla linguistica* (pp. 172–202). Milano: Feltrinelli (Discussione su "Linguistica e biologia", 308–319).

Prodi, G. (1984). La biologia come semiotica naturale. In: Bonfantini, Massimo A.; Ferraresi, Mauro (eds.), La ragione abduttiva. *Il Protagora, 24*(6), 85–104.

Prodi, G. (1986a). Semiosic competence, development of. In T. A. Sebeok (Ed.), *Encyclopedic dictionary of semiotics. Vol 2, N–Z, Approaches to Semiotics 73* (pp. 884–887). Berlin: Mouton de Gruyter.

Prodi, G. (1986b). Interview. In G. Marrone (Ed.), *Dove va la semiotica? Quaderni del Circolo Semiologico Siciliano 24*. Palermo: Circolo Semiologico Siciliano.

Prodi, G. (1987). *Curriculum vitae ed elenco delle pubblicazioni*. [62 pp. + 5 pp. with an additional list from 1988–1990. A copy in T. A. Sebeok memorial library, Department of Semiotics, University of Tartu.]

Prodi, G. (1988a [1977]). Material bases of signification. *Semiotica, 69*(3/4), 191–241.

Prodi, G. (1988b). *Teoria e metodo in biologia e medicina*. Bologna: Editrice CLUEB.

Prodi, G. (1988c). Signs and codes in immunology. In: Sercarz et al. 1988: 53–64.

Prodi, G. (1988d). La cultura come ermeneutica naturale. *Intersezioni, Rivista di storia delle idee* (Bologna: Società Editrice Il Mulino) *8*(1): 23–48. [English translation in Prodi 1989b]

Prodi, G. (1988e). La biologia come semiotica naturale. In M. Herzfeld & L. Melazzo (Eds.), *Semiotic theory and practice: Proceedings of the third international congress of the IASS Palermo, 1984* (Vol. 2, pp. 929–951). Berlin: Mouton de Gruyter.

Prodi, G. (1989a). L'interpretazione come cambiamento dell'interprete. *Versus, 52*(53), 21–24.

Prodi, G. (1989b). Culture as natural hermeneutics. In W. A. Koch (Ed.), *The nature of culture. Proceedings of the international and interdisciplinary symposium, October 7—11, 1986 in Bochum, Bochum Publications in Evolutionary Cultural Semiotics; BPX 12* (pp. 215–239). Bochum: Studienverlag Dr. Norbert Brockmeyer [Translation of Prodi 1988d].

Prodi, G. (1989c). Biology as natural semiotics. In W. A. Koch (Ed.), *For a semiotics of emotion, Bochumer Beiträge zur Semiotik; BBS 4* (pp. 93–110). Bochum: Brockmeyer.

Prodi, G. (1989d). Toward a biologically grounded ethics. *Alma Mater Studiorum, 2*(1), 65–73.

Prodi, G. (1994). Semiosic competence, development of. In T. A. Sebeok (Ed.), *Encyclopedic dictionary of semiotics. Vol 2, N–Z, Approaches to Semiotics 73* (2nd ed., pp. 884–887). Berlin: Mouton de Gruyter [Reprint of Prodi 1986].

Prodi, G. (2002). La biologia come semiotica naturale. *Athanor, 5*, 63–72.

Prodi, G. (2010a). Signs and codes in immunology (1988). In: Favareau 2010a: 328–335. [Reprint of Prodi 1988c.]

Prodi, G. (2010b). Semiosic competence, development of. In T. A. Sebeok & M. Danesi (Eds.), *Encyclopedic Dictionary of Semiotics. Vol 2, N–Z* (3rd ed., pp. 907–910). Berlin: De Gruyter Mouton [Reprint of Prodi 1986].

Robinson, A., & Southgate, C. (2010). A general definition of interpretation and its application to origin of life research. *Biology and Philosophy, 25*(2), 163–181.

Rodríguez Higuera, C. J. (2016). *The place of semantics in biosemiotics: Conceptualization of a minimal model of semiosic capabilities, dissertationes semioticae universitatis tartuensis 24*. Tartu: University of Tartu Press.

Rodríguez Higuera, C. J., & Kull, K. (2017). The biosemiotic glossary project: The semiotic threshold. *Biosemiotics, 10*(1), 109–126.

Sebeok, T. A. (1997a). The evolution of semiosis. In R. Posner, K. Robering, & T. A. Sebeok (Eds.), *Semiotics: A handbook on the sign-theoretic foundations of nature and culture* (Vol. 1, pp. 436–446). Berlin: Walter de Gruyter.

Sebeok, T. A. (1997b). Foreword. In R. Capozzi (Ed.), *Reading Eco: An anthology* (pp. xi–xvi). Bloomington: Indiana University Press.

Sebeok, T. A. (1998). The Estonian connection. *Sign Systems Studies, 26*, 20–41.

Sebeok, T. A. (2001). Biosemiotics: Its roots, proliferation, and prospects. *Semiotica, 134*(1/4), 61–78.

Sebeok, T. A. (2004). Origins: Semiosis the domain *vs.* Semiotics the field. In: Alač, Violi 2004: 83–104.

Sebeok, T. A., & Umiker-Sebeok, J. (Eds.). (1992). *Biosemiotics: Semiotic web 1991, Approaches to Semiotics 106*. Berlin: Mouton de Gruyter.

Sercarz, E. E., Celada, F., Michison, N. A., & Tada, T. (Eds.). (1988). *The semiotics of cellular communication in the immune system: Proceedings of the NATO advanced research workshop on the semiotics of cellular communication in the immune system held at Il Ciocco, Lucca, Italy, September 9–12, 1986, Nato ASI Series 23*. Berlin: Springer.

Sharov, A. (2013). Minimal mind. In L. Swan (Ed.), *Origins of mind, Biosemiotics 8* (pp. 343–360). Dordrecht: Springer.

Sharov, A. (2017). Molecular biocommunication. In R. Gordon & S. Joseph (Eds.), *Biocommunication: Sign-mediated Interactions Between Cells And Organisms* (pp. 3–35). New Jersay: World Scientific.

Sharov, A., & Vehkavaara, T. (2015). Protosemiosis: Agency with reduced representation capacity. *Biosemiotics, 8*(1), 103–123.

Smith, W. J. (1965). Message, meaning, and context in ethology. *The American Naturalist, 99*(908), 405–409.

Stjernfelt, F. (2014). *Natural Propositions: The actuality of Peirce's Doctrine of Dicisigns*. Boston: Docent Press.

Tønnessen, M. (2001). Outline of an Uexküllian bio-ontology. *Sign Systems Studies, 29*(2), 683–691.

von Uexküll, J. (1928). *Theoretische biologie* (2nd ed.). Berlin: Verlag von Julius Springer.

Weber, A. (2016). *Biopoetics: Towards a Biological Theory of Life-as-Meaning, Biosemiotics 14*. Berlin: Springer.

Bibliography

Prodi's Main Philosophical and Biosemiotic Works

In Italian

Prodi, G. (1974). *La scienza, il potere, la critica*. Bologna: Il Mulino.
Prodi, G. (1977). *Le basi materiali della significazione*. Milano: Bompiani.
Prodi, G. (1979). *Orizzonti della genetica*. Milano: Espresso Strumenti.
Prodi, G. (1982). *La storia naturale della logica*. Milano: Bompiani.
Prodi, G. (1983a). *L'uso estetico del linguaggio*. Bologna: Il Mulino.
Prodi, G. (1983b). Lingua e biologia; Discussione su 'Linguistica e biologia'. In C. Segre (Ed.), *Intorno alla linguistica*. Milano: Feltrinelli p. 172–202; p. 308–319.
Prodi G. (1984). La biologia come semiotica naturale. In: Bonfantini, Massimo A and Ferraresi, Mauro (eds.), *La ragione abduttiva*. Il Protagora 24(6): 85–104.
Prodi, G. (1985). L'ingegneria genetica fra biologia e filosofia. *Il Mulino, 299*, 420–470.
Prodi, G. (1986). Dove va la semiotica? *Quaderni del circolo semiologico siciliano, 24*, 121–132.
Prodi, G. (1987a). *Alla radice del comportamento morale*. Genova: Marietti.
Prodi, G. (1987b). *Gli artifici della ragione*. Milano: Edizioni del Sole 24 ore.
Prodi, G. (1988b). *Teoria e metodo in biologia e medicina*. Bologna: Editrice CLUEB.
Prodi, G. (1988c). La cultura come ermeneutica naturale. *Intersezioni, VIII*, 23–48.
Prodi, G. (1989). *L'individuo e la sua firma. Biologia e cambiamento antropologico*. Bologna: Il Mulino.

In English

Prodi, G. (1988a). Material bases of signification. *Semiotica, 69*(3–4), 191–241.
Prodi, G. (1988d). Signs and codes in immunology In: Sercarz, E. E., Celada, F, Mitchison, N. A., & Tada, T. (eds.), *The semiotics of cellular communication in the immuno system* (pp. 53–64). [NATO ASI Series, Series H: Cell Biology, vol. 23.] Berlin: Springer-Verlag (reprinted in: Favareau D. (ed.), *Essential readings in biosemiotics: Anthology and commentary* (pp. 323–336). Berlin: Springer 2010).

Prodi, G. (1989a). Culture as natural hermeneutics. In W. Koch (Ed.), *The nature of culture.*
    Proceedings of the international and interdisciplinary symposium, October 7—11, 1986 in
    Bochum (pp. 215–239). Bochum: Brockmeyer.
Prodi, G. (1989b). Biology as natural semiotics. In W. Koch (Ed.), *For a semiotics of emotion*
    (pp. 93–110). Bochum: Brockmeyer.
Prodi, G. (2010). Signs and codes in immunology. In D. Favareau (Ed.), *Essential readings in*
    biosemiotics: Anthology and commentary (pp. 323–336). Berlin: Springer (originariamente
    apparso in: Sercar E ed. The semiotics of cellular communication in the immune system,
    NATO ASI Series H, Cell Biology. Heidelberg).

Other Works Cited

Aboitiz, F. (Ed.). (2017). *A brain for speech. A view from evolutionary neuroanatomy.* New York:
    Springer.
Antonello, P. (2012). *Contro il materialismo. Le "due culture" in Italia: bilancio di un secolo.*
    Torino: Aragno.
Baker, L. (1989). *Saving belief. A critique of physicalism.* Princeton: Princeton University Press.
Baldwin, J. (1896). A new factor in evolution. *The American Naturalist, 30*(354), 441–451.
Barbieri, M. (2006). Life and semiosis: The real nature of information and meaning. *Semiotica,*
    158, 233–254.
Barbieri, M. (2009). A short history of biosemiotics. *Biosemiotics, 2,* 221–245.
Barbieri, M. (2014). From biosemiotics to code biology. *Biological Theory, 9*(2), 239–249.
Barbieri, M. (2015). *Code biology. A new science of life.* Dordrecht: Springer.
Barrow, J., & Tipler, F. (1988). *The anthropic cosmological principle.* Oxford: Oxford University
    Press.
Baslow, M. (2011). Biosemiosis and the cellular basis of mind. *Biosemiotics, 4*(1), 39–53.
Bekoff, M., & Pierce, J. (2009). *Wild justice: The moral lives of animals.* Chicago: University of
    Chicago Press.
Benveniste, E. (1971). *Problems in general linguistics.* Miami: Miami University Press.
Bolhuis, J., Brown, G., Richardson, R., & Laland, K. (2011). Darwin in mind: New opportunities
    for evolutionary psychology. *PLoS Biology, 9*(7), e1001109.
Bolhuis, J., Tattersall, I., Chomsky, N., & Berwick, R. (2014). How could language have evolved?
    PLoS Biology, 12(8), e1001934.
Boniolo, G., & De Anna, G. (Eds.). (2006). *Evolutionary ethics and contemporary biology.*
    Cambridge: Cambridge University Press.
Braitenberg, V. (1984). *I veicoli pensanti. Saggio di psicologia sintetica.* Milano: Garzanti.
Brenner, E., Stahlberg, R., Mancuso, S., Vivanco, J., Baluška, F., & Van Volkenburgh, E. (2006).
    Plant neurobiology: An integrated view of plant signaling. *Trends in Plant Science, 11*(8),
    413–419.
Brentari, C. (2015). *Jakob von Uexküll. The discovery of the umwelt between biosemiotics and*
    theoretical biology. Berlin: Springer.
Briscoe, T. (Ed.). (2002). *Linguistic evolution through language acquisition.* Cambridge:
    Cambridge University Press.
Buchanan, B. (2008). *Onto-Ethologies. The animal environments of Uexküll, Heidegger, Merleau-*
    Ponty, and Deleuze. New York: Suny Press.
Buyssens, E. (1943). *Le langage et le discours.* Bruxelles: BruxellesOffice de Publicité.
Caputo, C. (1990). Bio-logia vs semio-logia, La proposta di Giorgio Prodi. *Idee. Genesi del senso,*
    5(13–15), 183–188.

Cavalli-Sforza, L., Minch, E., & Mountain, J. (1992). Coevolution of genes and languages revisited. *Proceedings of the National Academy of Sciences of the United States of America, 89*(12), 5620–5624.

Chalmers, D. (1996). *The conscious mind. In search of the fundamental theory.* Oxford: Oxford University Press.

Chieco, B. L. (2011). Giorgio Prodi oncologo. *Belfagor, 66*(395), 609–612.

Chomsky, N. (1988). *Language and problems of knowledge. The Managua lectures.* Cambridge: The MIT Press.

Churchland, P. (1981). Eliminative materialism and the propositional attitudes. *The Journal of Philosophy, 78*(2), 67–90.

Cimatti, F. (1998). *Mente e linguaggio negli animali. Introduzione alla zoosemiotica cognitiva.* Roma: Carocci.

Cimatti, F. (2000a). The circular semiosis of Giorgio Prodi. *Sign Systems Studies, 28,* 351–379.

Cimatti, F. (2000b). *Nel segno del cerchio. L'ontologia semiotica di Giorgio Prodi.* Roma: Manifesto libri.

Cimatti, F. (2000c). *La scimmia che si parla. Linguaggio, autocoscienza e libertà nell'animale umano.* Torino: Bollati Boringhieri.

Cimatti, F. (2015). Italian philosophy of language. *Rivista Italiana di Filosofia del Linguaggio, 1,* 14–36.

D'Angelo, P. (2015). *Il problema Croce.* Macerata: Quodlibet.

Darwin, C. (2006). *The origins of species. By means of natural selection, or the preservation of Favoured races in the struggle for life.* Cambridge: Cambridge University Press.

Darwin, C. (2009). *The descent of man and selection in relation to sex.* Cambridge: Cambridge University Press.

De Nigris, F. (1981). Intervista a Giorgio Prodi. *Quarantacinque, 3*(10), 27.

de Saussure, F. (2011). *Course in general linguistics.* New York: Columbia University Press.

de Waal, F. (1996). *Good natured: The origins of right and wrong in humans and other animals.* Cambridge: Harvard University Press.

de Waal, F. (2008). Putting the altruism back into altruism: The evolution of empathy. *Annual Review of Psychology, 59,* 279–300.

Deacon, T. (1997). *The symbolic species, the co-evolution of language and the brain.* New York: Norton.

Deely, J., Petrilli, S., & Ponzio, A. (2005). *The semiotic animal.* New York: Legas.

Dennett, D. (1991). *Consciousness explained.* New York: Little, Brown and Company.

Dennett, D. (1995). *Darwin's dangerous idea. Evolution and the meanings of life.* New York: Simon & Schuster.

Dobzhansky, T. (1968). On some fundamental concepts of Darwinian biology. In T. Dobzhansky, M. Hecht, & W. Steere (Eds.), *Evolutionary biology* (pp. 1–34). New York: Springer.

Donati C. (1985) Intervista a Giorgio Prodi. *Il Resto del Carlino,* 23 novembre 1985: 16.

Dretske, F. (1995). *Naturalizing the mind.* Cambridge: MIT Press.

Eccles, J. (1994). *How the SELF controls its BRAIN.* Berlin: Springer.

Eco, U. (1973). *Segno.* Isedi: Milano.

Eco, U. (1976). *A theory of semiotics.* Bloomington: Indiana University Press.

Eco, U. (1989). Una sfida al mito delle due culture. *Il Belpaese, 7,* 166–168.

Eco, U. (1999). *Kant and the Platypus: Essays on language and cognition.* New York: Harcourt.

Eldredge, N., & Gould, S. (1972). Punctuated equilibria: An alternative to phyletic gradualism. In T. Schopf (Ed.), *Models in paleobiology* (pp. 82–115). San Francisco: Freeman.

Emmeche, C. (2004). Causal processes, semiosis, and consciousness. In J. Seibt (Ed.), *Process theories: Crossdisciplinary studies in dynamic categories* (pp. 313–336). Dordrecht: Kluwer.

Emmeche, C., & Kull, K. (Eds.). (2011). *Towards a semiotic biology: Life is the action of signs.* London: Imperial College Press.

Esposito, R. (2012). *Living thought. The origins and actuality of Italian philosophy.* Stanford: Stanford University Press.

Fadda, E. (2013). *Peirce*. Roma: Carocci.

Fadda E. (2014). Dalla parte di Cerbero. Peirce e la comunicazione. *Rivista Italiana di Filosofia del Linguaggio*, https://doi.org/10.4396/04SFL2014.

Favareau, D. (2010). Preface. A stroll through the worlds of science and signs. In D. Favareau (Ed.), *Essential readings in biosemiotics: Anthology and commentary* (pp. V–xi). Berlin: Springer.

Fehr, E., & Fischbacher, U. (2003). The nature of human altruism. *Nature, 425*, 785–791.

Fouts, R. (1997). *Next of kin*. New York: A Living Planet Press Book.

Frege, G. (1984). *Collected papers on mathematics, logic, and philosophy*. Oxford: Blackwell.

Freud, S. (2000). *Three essays on the theory of sexuality*. New York: Basis Books.

Gagneux, P., & Varki, A. (2001). Genetic differences between humans and great apes. *Molecular Phylogenetics and Evolution, 18*, 2–13.

Galef, B. (1988). Imitation in animals: History, definition, and interpretation of data from the psychological laboratory. In T. Zentall & B. Galef (Eds.), *Social learning: Psychological and biological perspectives* (pp. 3–28). Hillsdale: Erlbaum.

Garroni, E. (1986). *Senso e paradosso*. Bari: Laterza.

Garroni, E. (1992). *Estetica. Uno sguardo attraverso*. Milano: Garzanti.

Gibson, J. (1979). *The ecological approach to visual perception*. Boston: Houghton Mifflin.

Gómez, J. C., & Martín-Andrade, B. (2002). Possible precursor of pretend play in nonpretend actions of captive gorillas (*Gorilla gorilla*). In R. Mitchell (Ed.), *Pretending and imagination in animals and children* (pp. 255–268). Richmond: Eastern Kentucky University.

Gould, S. (2002). *The structure of evolutionary theory*. Cambridge: Harvard University Press.

Griffin, D., & Speck, B. (2004). New evidence of animal consciousness. *Animal Cognition, 7*(1), 5–18.

Guagnini, E. (2009). Letteratura come ipotesi e sperimentazione della realtà. Itinerario narrativo di Giorgio Prodi. In *L'opera narrativa di Giorgio Prodi* (pp. 7–17). Diabasis: Parma.

Harnad, S. (1990). The symbol grounding problem. *Physica D: Nonlinear Phenomena, 42*(1–3), 335–346.

Hauser, M. (2006). *Moral minds: How nature designed our universal sense of right and wrong*. New York: Ecco Press.

Hoffmeyer, J. (1996). *Signs of meaning in the universe*. Bloomington: Indiana University Press.

Hoffmeyer, J. (2010). The semiotics of nature: Code-duality. In D. Favareau (Ed.), *Essential readings in biosemiotics: Anthology and commentary* (pp. 583–628). Berlin: Springer.

Honig, W., & James, P. (Eds.). (1971). *Animal memory*. New York: Academic Press.

Horn, L. (1989). *A natural history of negation*. Chicago: The University of Chicago Press.

Humphrey, N. (1992). *A history of the mind: Evolution and the birth of consciousness*. New York: Simon & Schuster.

Hunt, G., Hopkins, M., & Lidgard, S. (2015). Simple versus complex models of trait evolution and stasis as a response to environmental change. *PNAS, 112*(16), 4885–4890.

Ingold, T. (2006). Against human nature. In N. Gontier, J. P. van Bendegem, & D. Aerts (Eds.), *Evolutionary epistemology, language and culture. A non-adaptationist, systems theoretical approach* (pp. 259–281). Berlin: Springer.

Kirshner, D., & Whitson, J. (Eds.). (1997). *Situated cognition: Social, semiotic, and psychological perspectives*. Mahwah: Erlbaum.

Knight, C., Studdert-Kennedy, M., & Hurford, J. (2000). *The evolutionary emergence of language*. Cambridge: Cambridge University Press.

Knobe, J., & Nichols, S. (Eds.). (2008). *Experimental philosophy*. Oxford: Oxford University Press.

Krampen, M. (1981). Phytosemiotics. *Semiotica, 36*(3/4), 187–209.

Kull, K. (1999). Umwelt and evolution: From Uexküll to Post-Darwinism. In E. Taborsky (Ed.), *Semiosis, Evolution, Energy* (pp. 53–70). Aachen: Shaker Verlag.

Kull, K. (2000). An introduction to phytosemiotics: semiotic botany and vegetative sign systems. *Sign Systems Studies, 28*, 326–350.

Kull, K., Deacon, T., Emmeche, C., Hoffmeyer, J., & Stjernfelt, F. (2009). Theses on biosemiotics: Prolegomena to a theoretical biology. *Biological Theory, 4*(2), 167–173.

Laland, K., & Galef, B. (Eds.). (2009). *The question of animal culture*. Cambridge: Harvard University Press.

Laland, K., Matthews, B., & Feldman. (2016). An introduction to niche construction theory. *Evolutionary Ecology, 30*, 191–202.

Lanyon, S. (2006). A saltationist approach for the evolution of human cognition and language. In A. Cangelosi, A. Smith, & K. Smith (Eds.), *The evolution of language* (pp. 176–183). Singapore: World Scientific.

Lavazzo, A., & Robinson, E. (Eds.). (2014). *Contemporary dualism: A defense*. London: Routledge.

Lavers, C. (2014). *The natural history of unicorns*. London: Granta Books.

Lenneberg, E. (1967). *Biological foundations of language*. New York: Wiley.

Lestel, D. (1995). *Paroles de singes. L'impossible dialogue homme-primate*. Paris: Éditions La découverte.

Levelt, W. (1989). *Speaking: From intention to articulation*. Cambridge: The MIT Press.

Lo, P. F. (1992). Le signe linguistique est-il à deux faces? Saussure et la topologie. *Cahiers Ferdinand de Saussure, 45*, 213–221.

Longo, G. (2011). Presentazione dell'Opera narrativa di Giorgio Prodi. *Intersezioni, 3*, 459–472.

Lorenz, K. (1981). *The foundations of ethology*. Berlin: Springer.

Lyn, H. (2017). The question of capacity: Why enculturated and trained animals have much to tell us about the evolution of language. *Psychonomic Bulletin & Review, 24*(1), 85–90.

Mainardi, D. (1973). *L'animale culturale*. Milano: Rizzoli.

Maran, T., Martinelli, D., & Turovski, A. (Eds.). (2011). *Readings in Zoosemiotics*. Berlin: De Gruyter Mouton.

Marks, J. (2015). *Tales of the ex-apes: How we think about human evolution*. Oakland: California University Press.

Martinelli D., Lehto O. eds. (2009). Zoosemiotics. *Sign Systems Studies 37*(3/4).

Mayr, E. (1982). *The growth of biological thought*. Cambridge: The Belknap Press of Harvard University Press.

Mazzoli, G., & Zucal, S. (Eds.). (1989). *Giorgio Prodi e l'avventura del pensare poliedrico*. Trento: Quaderni de "Il Margine".

McGinn, C. (1993). *Problems in philosophy: The limits of inquiry*. Oxford: Blackwell.

Mead, G. H. (1922). A behavioristic account of the significant symbol. *The Journal of Philosophy, 19*(6), 157–158.

Melandri, E. (2004). *La linea e il circolo*. Macerata: Quodlibet.

Mendelson, T., Fitzpatrick, C., Hauber, M., Pence, C., Rodriguez, R., Safran, R., Stern, C., & Stevens, J. (2016). Cognitive phenotypes and the evolution of animal decisions. *Trends in Ecology & Evolution, 31*(11), 850–859.

Millikan, R. (1984). *Language, thought and other biological categories*. Cambridge: The MIT Press.

Minsky, M. (1985). *The society of mind*. New York: Simon & Schuster.

Moore, G. (1959). *Principia Ethica*. Cambridge: Cambridge University Press.

Nesteruk, A. (2013). A "participatory universe" of J. A. Wheeler as an intentional. Correlate of embodied subjects and an example of purposiveness in physics. *Journal of Siberian Federal University. Humanities & Social Sciences, 6*(3), 415–437.

O'Connell, S. (1995). Empathy in chimpanzees: Evidence for theory of mind? *Primates, 36*(3), 397–410.

Ogden, C., & Richards, I. (1923). *The meaning of meaning*. New York: Hartcourt & Brace.

Pakendorf, B. (2014). Coevolution of languages and genes. *Current Opinion in Genetics & Development, 29*, 39–44.

Papineau, D. (2003). *Thinking about consciousness*. Oxford: Oxford University Press.

Parisi, D. (1999). *Mente. I nuovi modelli della Vita Artificiale*. Bologna: Il Mulino.

Parker, K. (1998). *The continuity of Peirce's thought*. Nashville: Vanderbildt University Press.

Peirce C.S. (1931–1958) *The collected papers of Charles Sanders Peirce*. Cambridge: The Belknap Press of Harvard.

Pepperberg, I. (2017). Animal language studies: What happened? *Psychonomic Bulletin & Review, 24*(1), 181–185.

Preston, S., & de Waal, F. (2002). Empathy: Its ultimate and proximate bases. *Behavioral and Brain Sciences, 25*(1), 1–20.

Prieto L. (1991) *Saggi di semiotica,vol. I, Sulla conoscenza*. Parma: Pratiche.

Rilling, J. (2006). Human and nonhuman primate brains: Are they allometrically scaled versions of the same design? *Evolutionary Anthropology, 15*(2), 65–77.

Rodríguez Higuera, C. J., & Kull, K. (2017). The biosemiotic glossary project: The semiotic threshold. *Biosemiotics, 10*(1), 109–126.

Roitblat, H. (2014). *Animal cognition*. New York: Psychology Press.

Rorty, R. (Ed.). (1967). *The linguistic turn: Recent essays in philosophical method*. Chicago: University of Chicago Press.

Rossi Landi, F. (1992). *Between sings and non-signs*. Amsterdam: Benjamins.

Ryle, G. (1949). *The concept of mind*. London: Hutchinson &.

Schino, G., & Aureli, F. (2010). The relative roles of kinship and reciprocity in explaining primate altruism. *Ecology Letters, 13*(1), 45–50.

Schrödinger, E. (2013). *What is life?* Cambridge: Cambridge University Press.

Sebeok, T. (1968). Zoosemiotics. *American Speech, 43*(2), 142–144.

Sebeok, T. (1986). *I think I am a verb. More contributions to the doctrine of signs*. New York: Springer.

Sebeok, T. (1998). The Estonian connection. *Sign Systems Studies, 26*, 20–41.

Sebeok, T. (2001). Biosemiotics: Its roots, proliferation, and prospects. *Semiotica, 134*(1/4), 61–78.

Shannon, C., & Weaver, W. (1949). *The mathematical theory of communication*. Chicago: University of Illinois Press.

Sheldon P. (2001) *Punctuated equilibrium and phyletic gradualism*. Encyclopedia of Life Sciences, Wiley www.els.net.

Sherwood, C., Subiaul, F., & Zawidzki, T. (2008). A natural history of the human mind: Tracing evolutionary changes in brain and cognition. *Journal of Anatomy, 212*(4), 426–454.

Singer, P. (1981). *The expanding circle: Ethics and sociobiology*. New York: Farrar.

Spinoza, B. (2002). *Complete works*. Indianapolis: Hackett.

Tamm, M., & Kull, K. (2016). Toward a reterritorialization of cultural theory: Estonian theory from Baer via Uexküll to Lotman. *History of Human Sciences, 29*(1), 75–98.

Tattersall, I. (2016). A tentative framework for the acquisition of language and modern human cognition. *Journal of Anthropological Sciences, 94*, 157–166.

Tennie, C., Call, J., & Tomasello, M. (2009). Ratcheting up the ratchet: On the evolution of cumulative culture. *Philosophical Transactions of the Royal Society B, 364*, 2405–2415.

Tomasello, M., & Call, J. (2004). The role of humans in the cognitive development of apes revisited. *Animal Cognition, 7*(4), 213–215.

Vermunt, M., et al. (2016). Epigenomic annotation of gene regulatory alterations during evolution of the primate brain. *Nature Neuroscience, 19*(3), 494–503.

von Uexküll, J. (1982). The theory of meaning. *Semiotica, 42*(1), 25–82.

von Uexküll, T. (1987). The sign theory of Jakob von Uexküll. In M. Krampen, K. Oehler, R. Posner, T. A. Sebeok, & T. von Uexküll (Eds.), *Classics of semiotics, Topics in Contemporary Semiotics* (pp. 147–179). Boston: Springer.

Vygotsky, L., & Luria, A. (1994). Tool and symbol in child development. In R. van der Veer & J. Valsiner (Eds.), *The Vygotsky reader* (pp. 99–174). Oxford: Blackwell.

West, D., & Anderson, M. (Eds.). (2016). *Consensus on Peirce's concept of habit: Before and beyond consciousness*. New York: Springer.

Wheeler, J. (1980). Beyond the black hole. In H. Wolf (Ed.), *Some strangeness in the proportion: A centennial symposium to celebrate the achievements of Albert Einstein* (pp. 341–375). Boston: Addison- Wesley.

Wilson, E. (1978). *On human nature*. Cambridge: Harvard University Press.

Wittgenstein, L. (1922). *Tractatus logico-philosophicus*. London: Kegan, Trench, Trubner.

Wittgenstein, L. (1993). Lecture on ethics. In J. Klagge & A. Nordmann (Eds.), *Philosophical occasions: 1912–1951* (pp. 36–44). Cambridge: Hackett.

Wittgenstein, L. (1998). *Notebooks 1914–1916*. Oxford: Blackwell.

Witzany, G. (2014). Pragmatic turn in biology: From biological molecules to genetic content operators. *World Journal of Biological Chemistry, 5*(3), 279–285.

Yasnitsky, A., & Van der Veer, R. (Eds.). (2016). *Revisionist revolution in Vygotsky studies: The state of the art*. New York: Routledge.

Zahavi, A., & Zahavi, A. (1997). *The handicap principle. A missing piece of Darwin's puzzle*. Oxford: Oxford University Press.

Zelenitsky, D., Therrien, F., Erickson, G., DeBuhr, C., Kobayashi, Y., Eberth, D., & Hadfield, F. (2012). Feathered non-avian dinosaurs from North America provide insight into wing origins. *Science, 338*(6106), 510–514.

Zorzella, C., & Cappi, M. (2012). L'uomo come specie di comunicazione. Bio-logica e strutture della significazione nella semiotica di Giorgio Prodi. In *Glossematica e semiotica: loro espansioni* (pp. 181–199). Padova: Zel Edizioni.

Index

© Springer Nature Switzerland AG 2018
F. Cimatti, *A Biosemiotic Ontology*, Biosemiotics 18,
https://doi.org/10.1007/978-3-319-97903-8

The manufacturer's authorised representative in the EU is Springer
Nature Customer Service Centre GmbH, Europaplatz 3, 69115 Heidelberg,
Germany. If you have any concerns regarding our products, please
contact ProductSafety@springernature.com

Printed and bound by CPI Group (UK) Ltd, Croydon, CR0 4YY
28/04/2026
02098475-0001